A WORLD BEYOND
PHYSICS

A WORLD BEYOND PHYSICS

The Emergence and Evolution of Life

Stuart A. Kauffman

OXFORD
UNIVERSITY PRESS

Oxford University Press is a department of the University of Oxford. It furthers
the University's objective of excellence in research, scholarship, and education
by publishing worldwide. Oxford is a registered trade mark of Oxford University
Press in the UK and certain other countries.

Published in the United States of America by Oxford University Press
198 Madison Avenue, New York, NY 10016, United States of America.

CIP data is on file at the Library of Congress
ISBN 978–0–19–087133–8

10

Printed by Sheridan Books, Inc., United States of America

CONTENTS

PROLOGUE

Classical physics, our gift from Newton, is our world written in the passive voice: rivers flow, rocks fall, planets orbit, stars arc in the space-time deformed by their masses. There are no doings, only happenings: myriad, miraculous, but brute.

I broach 78 years as I sit to write, having angled to the kitchen to pick out a nectarine to eat. Yesterday, I clambered aboard the "Poised Realm," my 22-foot boat, to skiff across to the Crane Dock on Orcas Island to drive to Eastsound, Washington, to buy the nectarine I just retrieved as an afternoon snack. My heart thumps a bit, my own human heart. Most of my readers have a human heart as well.

Just where did my human heart, the nectarine, my kitchen, the boat, and Eastsound come from since the brute happening of the Big Bang 13.7 billion years ago?

Since Newton, we have turned to physics to assess reality: what is REAL. But physics will not tell us whence we come, how arrived, why the human heart exists, nor why I can buy nectarines in Eastsound, let alone what "buying" is.

We will talk of these things, for there is more to know than we know and more to say than we can say.

We are in a world beyond physics.

We are in a world of living creatures that construct themselves. Yet we lack the concepts to say it. A tree, from a seed, builds itself, launches itself upward toward the sun. We see it and do not yet know what to say. A forest builds itself, rooted, branched, quiet, as if longing. Our biosphere too grows in diversity into what it can become and has done so for some 3.7 billion years. A giraffe? Who knew three billion years ago? None could have known. And nectarines: Who could then have said?

We estimate that 50 to 90 percent of the 10 to the power of 22 (10^{22}) stars in the known universe have planets cycling them. If, as I believe and will say, life is abundant, the universe is rife with becoming, based on physics but beyond any physics we know.

The concept of perhaps 10^{22} biospheres staggers me. Yes, we thrill at Hubble's image of billions of galaxies, some 10^{11} of them. But are there 10^{22} biospheres, ebullient like ours? Not "a world beyond physics," but "worlds beyond physics," as vast as the vastness of the physics we know, almost unknowable.

We miss in our science the idea of a system that constructs itself. I will introduce the requisite concept due to Maël Montévil and Mateo Mossio (2015) called "Constraint Closure". These young scientists have found a, or maybe "the," missing concept of biological organization. We will grow to understand it clearly and build on it. The ideas are a tiny bit complex, but not very. We will get to them. But for now we can think of constraint closure like this: it is a set of both constraints on the release of energy in non-equilibrium processes, and those processes, such that the system constructs its own constraints. This is an amazing idea. Cells do this, automobiles do not.

Living systems achieve this constraint closure and do what are called "thermodynamic work cycles" by which they can reproduce themselves. Living systems also exhibit Darwin's heritable variation, so can undergo his natural selection, hence evolve. I've written about that in some of my earlier books. But I was nagged by a feeling that there was something missing. With constraint closure a crucial puzzle piece is put into place.

But what evolves cannot be said ahead of time: what evolves emerges *unprestatably*—I know of no better word—and builds our biosphere of increasing complexity. We are its children: as are giraffes, nectarines, and sea cucumber.

Some years ago, at his 70th birthday fest, a physicist friend of mine smiled at the way biologists see the world. Were biologists with Galileo on the tower of Pisa, they would have dropped red stones, orange stones, pink stones, blue stones, green stones, and so on.

My physicist colleagues chuckled knowingly. Physicists seek to simplify to find laws, biologists to study how life became complex. So of course, the red stones were giraffes; the orange stones nectarines; the blue stones sea cucumbers; and the green stones, well, just us. The question is not whether the sea cucumber, giraffe, us, or the nectarine falls faster, but where did they come from in the very first place?

Physics won't say. No one knows.

There is a world beyond physics.

Darwin taught that new species drive wedges into the crowded floor of nature to make room for their own existence: yes, but no. Creatures, by existing, create the very conditions for other creatures to come into existence. Species constitute the very cracks in the floor of nature that constitute the niches for yet new species to come into existence, creating yet more cracks for still more species to spring forth.

The blossoming biosphere creates its own ever-new possibilities of becoming, yet more diverse and abundant.

The same holds, almost unnoted, for the exploding global economy. New goods create niches for yet further new goods: the invention of the World Wide Web created niches for selling on the Web, hence eBay and Amazon; which in turn created content on the Web, hence niches for search engines like Google; and for businesses that try to game, the search algorithms to sell more stuff. Or think of all the iPhone apps: and apps upon apps, like the ad blockers that remove the sales pitches from what Safari shows.

We stumble into the world we make possible as we lumber forward, with no or little insight or foreknowledge. I can go to Eastsound to buy nectarines.

We think that in physics—Special and General Relativity, Quantum Mechanics and Quantum Field Theory with the Standard Model—we will find the foundations from which we can derive the world, the ultimate becoming. We cannot. The ultimate may rest on the foundations, but it is not derivable from them. This ultimate, an unknowable unfolding, slips its foundational moorings and floats free. As Heraclitus said, the World Bubbles Forth.

Chapter 1

The World Is Not a Machine

Since the triumphs of Descartes, Newton, and Laplace and the birth of classical physics, we have come to regard physics as the answer to our questions about what reality "is." In that search, we have come to think of the world as a vast machine. This Newtonian fundamental framework is wonderfully extended by Special and General Relativity. Quantum Mechanics, and Quantum Field Theory, alter some of the basic deterministic aspects of classical physics but not the view of reality as an enormous "machine."

My thesis in this book is that, with respect to an evolving biosphere, ours and any in the universe, the "machine" thesis is wrong. Evolving life is not a machine. Elaborating how this is so will require some patience on all our parts. The consequences of the change in world view here proposed cannot be anticipated, but I hope they will include the realization that we are members of a living world of untellable creativity in its becoming. Along with that, I hope, will come a profound joy—an expanded awareness, a heightened appreciation, and a deepened sense of responsibility for the living world. Time will tell.

C. P. Snow wrote his famous essay, "The Two Cultures," decrying the split between the world of science and the world of art. Part of this split is the distinction between "mute" matter and the human imagination. But between these stands the evolving, living world, whether dumb of awareness or broadly conscious. I hope to show you that, unlike physics, where laws hold sway, no laws at all entail the becoming of the biosphere. No one knows or can know what shall become as the biosphere evolves and shapes its own future in ways we cannot state in advance. They are "unprestatable." This lawless emergence, contingent yet not random, bespeaks a place between mute matter and Shakespeare. Life itself spans between physics and art.

Please join me in exploring these issues, here hardly stated. There is much to do, more than this book can hope to accomplish. But I will try to give us a good start.

The Nonergodic Universe above the Level of Atoms

Has the universe made all the possible types of stable atoms? Yes. Bosons and fermions—the two broad kinds of particles physics knows—have glommed together in every conceivable combination to yield the hundred-odd elements that make matter. But will the universe make all possible complex things? No, not at all. Most complex things will never get to exist at all.

It is easy to see why: proteins are linear sequences of 20 kinds of amino acids—alanine, phenylalanine, lysine, tryptophan, and so forth. The specific sequence of these 20 amino acids along the "main chain" of a specific protein, linked by peptide bonds, defines the primary sequence of that protein. Thereafter the protein folds in complex ways to perform its functions in the cell.

A typical protein in humans is a linear sequence of some 300 amino acids. Some proteins are thousands of amino acids long.

How many possible proteins are there with only 200 amino acids in each? There are 20 choices at each position, so the total number of possible proteins of length 200 is 20 raised to the 200th power. That is about 10 raised to the 260th power. This is a hyper-astronomical number.

The next point is to see that the universe cannot have made more than a very small fraction of these possible proteins in the time since the Big Bang.

By our best reckoning, the age of the universe is about 13.7 billion years, which is about 10 to the 17th seconds. There are an estimated 10 to the 80th particles in the known universe. Quantum mechanics tell us that the shortest span of time in which anything can happen in the universe is the Planck time: 10 to the −43 seconds.

So if the 10 to the 80th particles in the universe were doing nothing since the Big Bang except making proteins in parallel at every tick of the Planck time clock, it would take 10 to the 39 power times the 13.7-billion-year actual history of the universe to make all possible proteins of the length 200 amino acids, *just once*. (In contrast, it may have taken only a few billion years to make all 20 amino acids.)

The universe, whatever might happen, can have made only a tiny fraction—1 over 10 to the 39th—of the possible proteins each consisting of 200 amino acids.

History enters when the space of the possible is vastly larger than what can become actual. For example, the evolution of life itself is a profoundly historical process. So too may be space chemistry and the formation of complex molecules. Thus, the becoming of the universe above the level of atoms is a historical process.

The physicist's phrase for this historicity is "nonergodic." "Ergodic" means, roughly, that the system visits all its possible states over some "reasonable" time period. The central example, from equilibrium statistical mechanics, is a liter volume of gas falling to equilibrium rapidly. The gas particles darting about in the bottle assume nearly every possible configuration before settling into the stablest possible state. But "nonergodic" means that a system does not visit all its possible states, like the amino acids that cannot make all possible proteins even after an astronomical number of repetitions of the 13.7-billion-year history of the universe.

If we ask whether the universe has created all stable atoms, the answer is yes. So the universe is roughly ergodic with respect to atoms, but it is not ergodic with respect to complex molecules. And the more complex the class of molecules, the more sparsely can that class have been sampled since the Big Bang. Consider proteins of length N = 1,2,3,4, . . . N + 1 amino acids. As N increases, the universe samples the possible sequences ever more sparsely. The universe can explore and surge upward in complexity indefinitely. In this sense, there is an indefinite "sink" upward in complexity. The universe can explore indefinitely vast realms.

Beyond the Second Law

The second law of thermodynamics states that disorder tends to increase. Disorder is measured as entropy. The canonical case again is that of a closed thermodynamic system of gas particles exploring every possible configuration in a liter container before settling into a state of equilibrium. It has reached what is called the most probable "macrostate"—the state of maximum entropy. The second law insists that entropy tends to increase as a system flows from less to more probable macrostates—like a steaming cup of coffee cooling to lukewarm and then cold, or like a cube of ice melting to a puddle.

But if everything moves inevitably toward a maximum state of entropy, then how can the universe—and especially the biosphere—be enormously complex? We really do not know. Part of the reason is that the universe itself is still on its way to equilibrium (the homogeneous murk that cosmologists call "heat death") and that the biosphere is not a closed system: the sun shines on us, providing energy to build complexity, forestalling entropy but only for a while.

A deeper part of the reason may be that the universe cannot exhaust complexity. Indefinite upward exploration into the vast possibilities of complexity occurs in terms of the complexity of space chemistry and the burgeoning diversity of the biosphere. Thus, we must ask how this indefinite complexity "sink" may bear on the emergent complexity of the universe. In particular, the biosphere has become complex with teeming diversity since its origin on the earth 3.7 billion years ago. Presumably this would be true of other living biospheres in the universe. Something in living biospheres surges "upward" in diversity and complexity. But how and why does it?

I hope to show you at least part of the source of this surging: a non-equilibrium companion of the famous second law, a principle that would help explain how the biosphere today can be far more complex than it was 4 billion years ago. Space chemistry shows this surging upward in complexity. After the Big Bang, the stable elements were created. The Murchison meteorite, formed some 5 billion years ago, has on the order of 14,000 types of organic molecules, cooked up from the elements of carbon, hydrogen, nitrogen, oxygen, phosphorus, and sulfur. The evolving biosphere manifests this surging upward in complexity, from protocells 3.7 billion years ago to the millions of species now. We seek nothing less than an understanding of where this order came from. The order is historically contingent, yet not fully random. Witness the order among the higher taxa as life explores the vast diversity of Darwin's "forms most beautiful."

The biosphere literally constructs itself and does so into a biosphere of increasing diversity. Again, How and why is this? Remarkably, the answer may be "Because the living world *can* become more diverse and complex and in an ongoing way *creates its own potential to do so*." That requires harnessing the release of energy to build order faster than that order can be dissipated by the second law of thermodynamics. As we will see, Montévil and Mossios's beautiful theory of Constraint Closure and thermodynamic work cycles figure prominently in our new story.

Why Do Human Hearts Exist?

Among the universe's endlessly unspooling complexities is the human heart. The universe in its lifetime can generate but a tiny fraction of all the possible proteins and even a tinier fraction of the tissues that are made from proteins and, in turn, make up the organs that we call hearts. So here is the question: Why *do* human hearts exist in this nonergodic universe that exists above the level of atoms?

Roughly, human hearts exist because they pump blood and thus were of selective advantage in our vertebrate ancestors and were inherited by us.

In short, Darwin gives us part of the answer: hearts help us survive, and so are selected. But Darwin did not realize that he was also giving a deeper account of why hearts exist at all: given ongoing, evolving life, with reproduction and heritable variation, if an organ arose with even a glimmering of the functional capacity to pump blood, that happy accident could be selected in organisms just a bit too large for simple diffusion to transport needed oxygen to every one of their cells. In short, *hearts exist in the nonergodic universe above the level of atoms by virtue of their functional role in abetting the survival of living, evolving organisms having such hearts.*

As they replicate, organisms propagate their organization of process—the way all of these pieces fit and work together. Organs are parts of that organization and exist *for and by means of the whole.* Hearts exist, in other words, because life exists. Moreover, as we will see, life creates the expanding space of possibilities into which it evolves in the nonergodic universe above the level of atoms.

That is the first major conclusion of this book: for complex things, getting to exist at all in the nonergodic universe above the level of atoms needs explanation, and the answer is as simple as it is profound. Hearts exist by *virtue* of their functional role in sustaining the existence, hence the evolving future, of living organisms with such hearts. Organisms propagate above the level of atoms, and so, with them, do their sustaining organs. Hearts exist in the nonergodic universe above the level of atoms because organisms need functioning hearts to exist and proliferate. As Kantian wholes, organisms carry along with them their sustaining parts. Organisms with hearts exist, so hearts exist.

Why do eyes exist, noses exist, kidneys exist, tentacles with suckers on them exist, sex exist, parental care exist, and the giraffe's long neck exist? The answer is the same: by virtue of the roles these organs and processes play in abetting the survival of the evolving, persistently living organisms having these organs and properties. They too exist for and by means of the whole.

All these aspects of our universe exist on a single blue-dot planet. If life is abundant among the estimated 10 to the 22nd

solar systems in the universe, what myriad of complex things—
unpredictable and perhaps unthinkable—lies in the indefinite
reach upward in complexity, ever further above atoms?

What Is an Organism?

Long before Darwin, Immanuel Kant understood this: "An organized being then, has the property that the parts exist for and
by means of the whole." Call this a "Kantian whole." The heart
exists for and by means of the whole organism of which it is a
functioning part. Humans are Kantian wholes.

A simple example of a Kantian whole is shown in Figure 1.1.
This is a hypothetical example of what I call a "collectively autocatalytic set." It consists of polymers, like the small proteins
called peptides. This system will be of central interest to us in
this book. It begins with simple "food molecules," single building
blocks we will call A and B (the monomers); and the four possible dimers, AA, AB, BA, and BB, all of which are supplied from
the outside. Then there are longer polymers, such as ABBA and
BAB, formed from this food set through reactions combining two
polymers end to end to create a longer polymer, or breaking a
longer polymer into two fragments. But here is the important
idea: the reactions forming these longer products are catalyzed
by the very polymers that make up the system. The system is collectively autocatalytic.

(A simpler example consists of two small polymers, AB and
BA, each formed by a reaction linking A and B. Here, AB catalyzes
the reaction forming BA, and BA catalyzes the reaction forming
AB. The set is collectively autocatalytic.)

In a set, such as that in Figure 1.1, no polymer catalyzes its
own formation; rather, the set as a whole catalyzes its joint formation. If one considers catalyzing a reaction a catalytic task, all

THE ORIGINS OF LIFE: A NEW VIEW

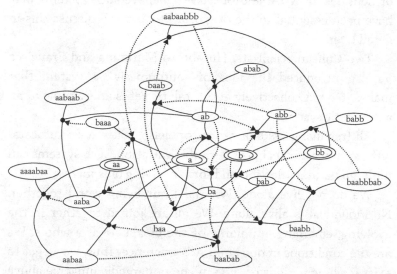

FIGURE 1.1 Collectively autocatalytic set. Symbol strings are molecules. Dots are reactions. Black arrows go from substrate molecules to reactions to product molecules. Dotted arrows from molecules to reactions show which molecule catalyzes which reaction. Double circles are exogenous food set. The function of a peptide or RNA is catalyzing the reaction forming the next peptide or RNA, not jiggling water in a petri plate.

the tasks are jointly realized in a kind of "catalytic task closure." Such a system is a "whole" and more than the sum of its parts. The closure of mutual catalysis is not seen in any of the parts alone. Rather, the closure is a collective property.

This system can literally build itself and reproduce! It is a Kantian whole, with parts that exist for and by means of the whole. It shall be my central model for the origin and even the character of life.

Collectively autocatalytic sets emerge spontaneously in sufficiently diverse chemical soups. Such systems exist, made of peptides, of RNA, and of DNA. I believe such systems may have been essential to the origin of life and will discuss this in detail later.

Two Chilean scientists, Humberto Maturana and Francisco Varela, introduced the idea of "autopoiesis," a system that makes itself. A collectively autocatalytic set is an example of an autopoietic system.

All free living systems are autopoietic, collectively autocatalytic, systems. If capable of heritable variation, such systems can undergo natural selection and form evolving biospheres.

We do not live alone. We make our living world together. No individual is alive alone. We all are joined together in the evolving, emerging, unfolding of the biosphere as a whole. We are the conditions of one another's existence. Thus, we all get to exist for long periods of time in the nonergodic universe above the level of atoms. Our biosphere has propagated stably for some 3.7 billion years.

These issues take us beyond our physics-based view of the world. Brilliant physicist Stephen Weinberg voiced the physicist thoughts: (1) the explanatory arrows always point downward from social systems to people to organs to cells, to biochemistry, to chemistry, and finally to physics; (2) the more we know of the universe, he wrote, the more meaningless it appears.

Yes, but we loudly say NO to these claims. We begin to see in this book that what gets to exist in the nonergodic biosphere above the level of atoms (hearts, sight, smell) exists *by virtue* of the functional roles these systems and subsystems play in abetting the survival and further evolution of the organisms of which they are parts. Hearing arose by co-opting the evolution of jawbones that were sensitive to vibration in an early fish and became the incus, males, and stapes of our middle ears. No one could have said, 3 billion years ago, that hearing would evolve. We could not have prestated that evolutionary emergence. But middle ear bones now exist in the nonergodic universe above the level of atoms by virtue of their emerged functional role in the survival and evolution of organisms having hearing. The explanatory arrows do *not* point downward from hearing to physics but upward to the selection of organs that abet hearing. That selection acted at the level of whole organisms as hearing evolved. That is why such organs exist in the universe, and Weinberg is just wrong.

I will return later to our incapacity to prestate the emergence of hearing, for from this it will follow that no laws at all "entail" the evolution of the biosphere, and that reductionism, Weinberg's dream of a final theory, is false.

The World as Machine

Prior to Descartes and Newton, the Western mind saw a Cosmos, an organic whole of which we were members. This was the view of the Church. With his res cogitans, Descartes reserved *mind* stuff for humans. The rest of the world, including our bodies and all animals and plants, were res extensa, extended stuff, mechanism. With the Principia of Newton, Aristotle's four causes—formal, final, efficient, and material—dwindled to a mathematized version of efficient cause: Newton's differential and integral calculus as captured in his three laws of motion and universal gravitation. Laplace's demon, knowing the positions and momenta of all the particles in the universe, could calculate its entire past and future. The world became a huge machine in classical physics, with its honest orbits. Modern reductionism was born. The theistic God retreated to a deistic God who set the universe up, selected initial conditions, and let Newton's laws take over. This God could no longer act in the world to create miracles. The struggle between science and religion flowered, and the rebellion of the

Romantics followed. It was "science with its rule and line," wrote Keats dismissively.

Weinberg is of this tradition. The scientific world is a machine and quite without meaning: a blight on Shakespeare and your natterings.

How overbold this is! The issues here have to do with including the towering problems of consciousness and agency, both vacant from the machine picture; how overbold indeed.

One thing missing in the world according to physics is the crucial idea of agency, which will concern us in a later chapter. Given agency, meaning exists in the universe, Weinberg or not. We are agents playing our complex, filigreed games with one another. Rocks play no games. Then, what must a system be to be an agent? What must systems be to evolve complex interwoven games of life with one another? This complexity is part of the complexity of the universe.

But leave out for now the magisterial issue of consciousness. Even if the biosphere were organisms dumb of consciousness, evolution would in no way be a world machine. In the nonergodic universe above the level of atoms, the biotic world surges beyond our saying, beyond equations and calculation in Laplace's style, beyond Keats's rued rule and line, and into the exploding adjacent possibilities life itself creates. The evolving biosphere becomes "sucked" into the very opportunities to explore untellable reaches of complexity, of never before seen organizations of matter and energy: life evolving. This evolution of the biosphere is an organic "whole." Its members jointly create the very pathways of further becoming what the biosphere as a whole shall become, derived of a past no less mysterious in its sinuous emergence. This living whole world is the Cosmos we lost with Descartes.

It will take the rest of this book to make good on the promise of these few paragraphs.

Chapter 2

The Function of Function

Perhaps the deepest, most troubling question about our strange and wonderful existence is this: How did the universe get from matter to mattering? In the meaningless, numb universe of Weinberg, where does mattering come from? Rocks are matter, but nothing matters to rocks. But, without attributing consciousness to a bacterium, glucose matters to a glucose-eating bacterium. What must a system be such that mattering can arise since the brute Big Bang?

Buried in our inquiry are questions that take us beyond physics—if the questions themselves are legitimate. We say, for example, "The function of the heart is to pump blood." But what then is a "function?" After all, pumping blood is just a "numb" causal consequence of the heart. But the heart also makes heart sounds and jiggles water in the pericardial sac. These too are causal consequences. But they are not functions. In short, functions are subsets of the causal consequences of parts of organisms. But how do we know which ones?

The issue is central to the question of reducing biology to physics. "Functions" in the biological sense, do not exist in physics. Consider a rubber ball. It is round, elastic, can spin on its axis, and it can bounce. But physics cannot say the function of the ball is to bounce. Nor can physics say the function of a river is to flow. Thus, if functions are a legitimate part of biology, then biology cannot be reduced to physics.

Here is an answer: the human heart, as we've seen, is part of a human Kantian whole; and of all the heart's causal consequences, the one that sustains the whole is pumping blood and not making thumping sounds, being red, jiggling water in the pericardial sac, and so forth. So pumping blood is its *function*—a *subset* of its causal consequences. Thus both the heart and the organism get to exist and persist in the nonergodic universe above the level of atoms.

More generally, to be a function something must abet the survival of a Kantian whole—like us or a fruit fly or any living thing.

Think again of our self-sustaining, self-creating autocatalytic set of peptides: the function of a peptide is to catalyze the formation of some other peptide and not to jiggle the water in the petri plate. Again, the peptide is part of a Kantian whole, the autocatalytic set, and its function is that subset of its causal consequences that help sustain that whole.

Two large conclusions follow. (1) We can justify the concept "function" in biology because things with functions, for example, hearts, get to exist in the nonergodic universe above the level of atoms by virtue of their role in living organisms—that is, as Kantian wholes that themselves propagate above the level of atoms. So functions are indeed a legitimate scientific concept; (2) the function of a part is typically a subset of its causal consequences, pumping blood and not jiggling water in pericardial sacs.

How different this is from physics. When a stream flows over rocks as it tumbles to the sea, the physicist can describe what happens but can pick out no subset of the happenings as a function. But the function of a peptide in an autocatalytic set, a Kantian whole, is its role in sustaining the catalytic, functional closure of the whole.

To a physicist the pumping, jiggling, shininess, and so forth of the heart are all on equal footing. None of them "matter."

We will see in this book that what "gets to exist" in the nonergodic universe above the level of atoms and in the evolving biosphere includes ever new, unprestatable "functions" that only come to exist because they abet the survival of the organisms having those functions, for example eyes and sight.

Thus, we are beyond physics for a second reason: physics, in principle, cannot predict these unprestatable new functions, such as hearing and middle ear bones, which come to exist. And so, again, biology cannot be reduced to physics.

The burgeoning diversity that is the biosphere surging upward in complexity for the past 3.7 billion years is surely based on physics, but it flowers to a realm beyond.

Chapter 3

Propagating Organization

Somehow, since the Big Bang, life emerged from non-life in ways we struggle to understand. Out my window on Crane Island in Puget Sound are deer, salmon, eagles, the occasional orcas, seals, herons, fir, and madroña. All of them flourish. Somehow it all appeared on earth starting some 3.7 billion years ago. Most scientists think, as I do, that life arose here and was not seeded from another world. Such a notion, panspermia, could be right, but it does not explain where life came from in the first place in the wide universe. We will discuss theories of the origin of life— whether on earth or elsewhere—in the following. Some of these, based on the spontaneous emergence of collectively autocatalytic sets, tie directly to the themes introduced in this chapter.

Life is transformative in the becoming of the non-living universe. Diversity exploded. The evolution of the biosphere has seen the sprawling emergence of the living diversity out your window, near and far. "Forms most beautiful," wrote Darwin. How in our world has this diversity come to exist and how has life propagated stably while it has diversified for 3.7 billion years? Yes, there is Darwin's heritable variation and natural selection, but whence arrives that which can undergo heritable variation and natural selection? Whence the arrival of the fitter (which was never answered by Darwin)? And before that, whence arrives life? And, once arrived, how does life propagate its organization of process?

Drawing on my own book, *Investigations* (Kauffman 2000), we broach these topics in the next few chapters. Life propagates organization that somehow links matter and energy in new ways to reproduce itself and, literally, to construct itself. The seed of the tree constructs, from inside itself, the tree it becomes. How so? The tree spawns offspring that have evolved to new kinds of trees for the past several hundred million years. How so? Yes, we know DNA, RNA and proteins, the double helix, the genetic code, central dogma, and all that; but these are not sufficient answers. It takes a whole cell to build a whole cell, thence to build an organism that, over generations of spawning, yields evolving populations of diversifying organisms. What is this organization that propagates? What is organization—and the ability to organize—that propagates?

Life somehow partially collaborates with and yet beats the second law of thermodynamics with its insistence that disorder—entropy—must inevitably increase in closed thermodynamic systems. How does life evade, but not avoid, that law?

A small part of the answer is that all living systems are *open* thermodynamic systems, taking in matter and energy. They are, in other words, displaced from equilibrium—that most probable state of maximum entropy to which a jar of gas molecules ultimately settles. As many, including Prigogine, have shown (Prigogine and Nicolis 1977), such systems can "eat" the order in their environment, such as gradients, and build order. Non-living systems, like whirlpools and Benard Cells, in which convective flow patterns emerge spontaneously in plates of viscous fluid slowly heated from below, show that pattern can emerge in such systems while displaced from equilibrium. Prigogine called these "dissipative structures" because they dissipate free energy.

In his famous book, *What Is Life?* (Schrödinger 1944), Schrodinger said life feeds on "negentropy," order in the environment that somehow is then converted into order in living systems.

Living systems propagate their organization. But what is this "propagating order?" What is the foundation of biological organization? Can we define this fundamental phenomenon?

As I have hinted, two young scientists, Maël Montévil and Matteo Mossio, may have recently found the essential missing concept, which they call Constraint Closure (Montévil and Mossio 2015). In this chapter, I hope to build toward their fine concept, then later build on it.

Work

We begin with the concept of work, another of those notions that seem so simple until you begin taking it apart. What is work? If one asks a physicist, work is force acting though a distance. So, I accelerate a hockey puck. The total acceleration of the mass is the amount of work done.

But already there is the start of a mystery: What or who picked the specific direction in which the puck was accelerated, say to the Northeast? The "amount of work done" does not specify this issue. It seems that for work to be done, something specific must happen, the puck must be accelerated to the Northeast and not in all directions on the frozen lake simultaneously.

Where does the specificity come from? Peter Atkins (1984) takes the next major step: work, says Atkins, is a thing! "Work is the constrained release of energy into a few degrees of freedom." It will take us a few moments to absorb this.

Think of a cylinder and piston, with the "working gas" con-
fined to the space between the cylinder and the piston. The
expanding gas does work on the piston to move it in the cylinder.
This is the constrained release of energy into a few degrees of
freedom.

For a physicist, "degrees of freedom" mean roughly what is
now possible. In the absence of the cylinder, the hot gas would
expand in all directions. No work would be done. In its presence,
the gas expands only along the cylinder, thrusting the piston.
And so work is done.

Boundary Conditions, Work, and Entropy

A physicist studying the system would impose fixed boundary
conditions for the cylinder and moving boundary conditions
for the piston. The fixed boundary condition specifies the loca-
tion of the cylinder. The moving boundary condition specifies
the positions of the moving piston inside the cylinder. Then the
physicist would solve for the work done in the system during
the process of the constrained release of energy as the gas drives
the piston down the cylinder.

Recall that from Newton, we have laws of motion, in the
form of differential equations and then initial and boundary
conditions. Seven billiard balls rolling on a billiard table are an
example. The initial conditions are the positions and momenta
of the balls; the boundary conditions are the shape of the table.
The boundary conditions are required to integrate the equations
of motion to obtain the work done.

What Atkins is telling us is that without the boundary
conditions that serve as constraints on the release of energy in a
non-equilibrium process, no work is done.

But there is more to consider here. During the expansion of the gas, as work is done, entropy increases but in a very specific way. The increase in entropy would be greater were the cylinder not there at all and the gas expanded in all spatial directions, that is, into all degrees of freedom, the full space of possibilities, everywhere. But the boundary conditions constrain the release of energy into only a few degrees of freedom, and only then is work done. As a result, the increase in entropy is less than were the constraints not there. The constraints, in other words, channel the release of energy into work, not just entropy increase. So here is another key concept: this channeling of work is part of how life "beats" the second law. Due to constraints, entropy still increases, but more slowly. This will be part of the answer, drawing on the notion of Constraint Closure, of how life surges upward in complexity and spreads this order despite the second law.

The Constraint Work Cycle

The physicist cheats when he or she just puts in boundary conditions for the cylinder and piston and leaves it at that. After all, just where did the cylinder come from since the Big Bang? Well, it took work to construct the cylinder that then serves as a constraint on the release of energy. It took work to construct the piston. It took work to assemble the piston inside the cylinder and arrange for the gas to be at the head of the cylinder. Physicists ignore this when they merely impose boundary conditions with no consideration of from whence they came. A locomotive is a large machine with many constraints on the release of energy. It takes work to build a locomotive.

It may not always take work for constraints to come into existence. Hot molten rock can congeal into tubes that constrain the flow of still molten lava. But living cells, as we will see, really do work to construct constraints channeling their own release of energy that constitutes further work

So surely, no constraints, no work. And often, no work, no constraints.

Call this the Constraint Work cycle.

We will see how living cells do work to construct the very constraints on the release of energy in non-equilibrium processes, which release then constitutes more work. We will build toward the idea of constraint closure.

So more! It takes constraints on the release of energy to get work—and the work done can construct yet more constraints!

Still there is more. These newly constructed constraints can then constrain yet further releases of energy that then constitute yet further work that can construct yet further constraints, and so forth and so forth. And so order can self-propagate!

Our machines do not do this. An automobile constrains the motion of many parts but does not construct new constraints. Life does.

As we soon will see, this propagating work and constraint construction can loop back to close on itself! Thus, a set of constraints on a set of non-equilibrium processes can achieve a work task closure that constructs the very same set of constraints. The constraints do work tasks that construct the same constraints, or boundary conditions.

The system can literally build itself! This is the amazing concept of Constraint Closure of Montévil and Mossio (Montévil and Mossio 2015).

Later we will see that collectively, autocatalytic sets achieve just this closure of constraints.

We approach this now.

Nonpropagating and Propagating Work

In Figure 3.1, I show a cannon and a cannon ball. The powder explodes with the release of energy constrained by the cannon and does work on the cannon ball, which is thus fired into the air. We see the cannon ball hitting the ground and making a hole with hot dirt—the leftover energy from the flight. The explosion is an exergonic, or spontaneous, process. It releases energy. The moving of the cannon ball is an endergonic, or nonspontaneous, process. It absorbs energy. The firing of the cannon ball comes with the release of energy, and the digging of the hole requires an input of energy.

Figure 3.2 shows a formalism I now introduce from Montévil and Mossio. C_i is a constraint on the release of energy. Here it is the cannon. A black arrow extends from Ci down to the @ sign in a non-equilibrium process, A----@--→ B. This process is the non-equilibrium explosion of the powder, and the @ sign signifies the "constraint" on the release of that energy due to the cannon, C_i. The constraint "acts on" the non-equilibrium process such that work is done.

Nonpropagating Work

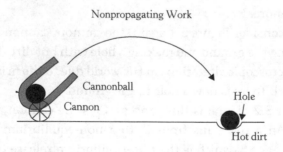

FIGURE 3.1 Cannon firing cannon ball, hits ground, hole, hot dirt = Nonpropagating work. From Kauffman, *Investigations* (Oxford University Press, 2000).

The cannon is the constraint on the release of powder explosion energy into a few degrees of freedom to do work to fire cannon ball out cannon.

C_i = Cannon constraint on release of energy

↓

A -- @ -- → B = constrained non-equilibrium process fires cannon ball and does work on cannon ball, an endergonic process.

But it took WORK to construct the cannon, and cannon ball, place powder inside and put cannon ball inside the cannon!

No constraints, no work. No work, no constraints. The W-C cycle!

FIGURE 3.2 The constrained release of energy does work

But now suppose that rather than hitting the ground and leaving a hole with hot dirt something simpler happens: the cannon ball hits a large hard steel plate and rolls to a stop. The collision sets up vibrations in the plate, which are dissipated as heat. Nothing has been made, not even a hole in the ground. No macroscopic alterations in the world have occurred beyond the firing of the cannon ball.

Call this nonpropagating work. It completes a task and does nothing more.

Now consider Figure 3.1 again, the cannon, cannon ball, and ball hitting the ground and making a hole with hot dirt. There are a few macroscopic alterations in the world due to the firing of the canon ball. There is now a hole in the ground.

Figure 3.2 diagrams this process. C_i is the cannon as a constraint. An arrow runs from C_i to a non-equilibrium process, A. . . . @. . . .->B, which is the non-equilibrium release of energy as the powder explodes and fires the cannonball. Due to the constraint, @, afforded by the cannon, the exploding power does work on the cannon ball.

Figure 3.3 is my own invention. The same cannon fires the very same cannon ball that strikes a paddle wheel I constructed that straddles a well I dug. The ball makes the wheel spin, thereby winding up a red rope I attached to a pail of water down the well.

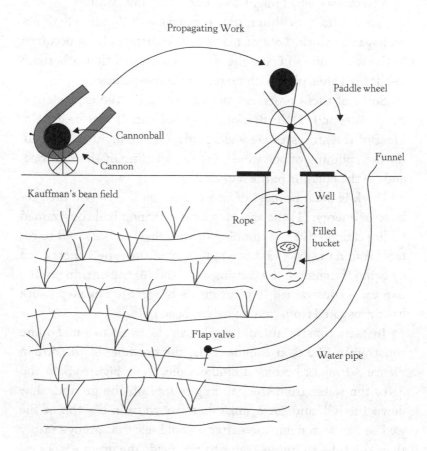

FIGURE 3.3 Propagating work. Cannon fires the same cannon ball that hits the paddle wheel doing work to make it spin, winding up a rope tied to an axle, and thereby doing work to wind up rope tied to the water-filled bucket in the well; that bucket tips over the axle and spills water into a funnel down a pipe to water my bean field. From Kauffman, *Investigations* (Oxford University Press 2000).

The winding of the rope pulls the pail up, it tips over the axle of the paddle wheel, and drops the water into a tube running down toward my bean field. The flowing water opens the flap valve at the bottom of the tube, thereby watering my crop.

You can see why I might be proud of my invention.

Beyond its agricultural use, the device in Figure 3.3 shows propagating work. Lots of macroscopic changes have occurred in the world due to firing the same cannon ball that otherwise might hit a steel plate with no further consequences.

Some of these processes are exergonic: (1) the explosion of the powder and (2) the flow of the spilled water to my bean field. Most of the processes are endergonic: (1) the flight of the ball, (2) the spinning of the wheel, (3) the winding of the red rope, and (4) the opening of the flap valve.

Work is done in most of these cases by the constrained release of energy: (1) the shooting of the cannon ball constrained by the cannon; (2) the spinning of the wheel constrained to rotate around the axle; (3) the winding up of the rope, constrained by being fastened to the turning axle; and (4) the opening of the flap valve constrained to pivot on its hinge. Step by step, work has propagated from cannon ball to bean field.

In fact, constraints and work can do work to build more constraints! The next time it rains, the hot hole in the dirt in Figure 3.1 might become a mud puddle. Or in Figures 3.4a and 3.4b, the water from the pail might spill on the ground, flow down the hill, and cut a small dirt groove from the top of the well to my bean field. Thereafter, I could use that groove rather than the tube to funnel water to the field. The groove is a new boundary condition.

In general, our familiar machines do propagating work. In a car, the gas explodes, pistons move in cylinders, the crank shaft turns, and the wheels turn. But the work done here constructs no new constraints or boundary conditions.

PROPAGATING WORK

C_i C_j (C_i = cannon; C_j = paddle wheel)

A----@--->B-----@--->C   (B = fired ball, C spinning wheel) Note: work is done on cannon ball by exploding powder and on paddle wheel hit by cannon ball

PROPAGATING WORK CONSTRUCTS A CONSTRAINT!, C_k

C_i C_j (C_i = cannon; C_j = paddle wheel)

A----@--->B-----@--->C_k (C_k is a new constraint. The water dumped from pail cuts ditch in hillside down to my bean field, can be used rather than the tube.)

FIGURES 3.4 Propagating work can construct new constraints

One more point before we head toward the first crescendo. After my bean field is watered, the cannon ball lies in the bushes somewhere and the pail lies beside the well. Can I add powder to the cannon and re-water my bean field? No. I have to find the cannon ball and replace it in the cannon and put the pail back down the well. In short, I have to complete what is called a thermodynamic work cycle. Hold onto that idea, for we shall quickly need it again.

Constraint Closure—and More

We come finally to the constraint closure of Montévil and Mossio. This is shown in Figure 3.5. The brilliant idea is now simple: work propagating through a linked set of constraints in a set of one or more non-equilibrium processes can do work to construct more constraints. Therefore, if this linked set of processes closes in on itself, the system can build the very same set of constraints used to constrain the release of energy by which it does work. The system can literally build itself including its constraints. This is constraint closure.

In Figure 3.5, I show a simple case. There are three non-equilibrium processes: (1) A----@--→C_k; (2) D----@--→C_L; and (3) G----@--→C_i. There are three constraints, C_i, C_k, and C_L: (1) C_i constrains the first process, showed by the arrow from C_i hitting the @ in the arrow for this process; (2) C_k constrains the second process; and (3) C_L constrains the third process. But process 3 makes the very first constraint, C_i! The set of propagating work constructs the exact same set of constraints, which themselves constrain the release of energy such that work is done in the first place.

Process 1 makes constraint 2, process 2 makes constraint 3, and process 3 makes constraint 1. This is a system where the set

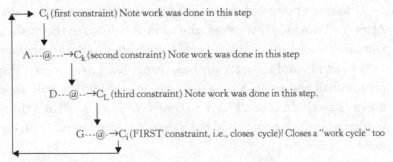

Constraint closure systems couple a set of non-equilibrium processes and the constraints on the releases of energy to do work that constructs the very same constraints, or boundary condition on the same non-equilibrium processes. This is a non-equilibrium self-constructing system that does a thermodynamic work cycle to construct and assemble its own parts into a working "whole"! IT CAN REPRODUCE ITSELF.

THIS IS A "MACHINE" THAT DOES WORK CYCLE TO BUILD AND ASSEMBLE ITS OWN WORKING PARTS! CARS DO NOT DO THIS! REPRODUCING CELLS DO THIS!

FIGURE 3.5 Montévil and Mossio's constraint closure: open non-equilibrium system

of constraints on the non-equilibrium processes harness each of these processes to do work to construct the very same set of constraints! This achieves constraint closure.

Two Closures

These systems actually show two different kinds of closures. First as Montévil and Mossio point out, there is constraint closure. The same system does work to construct the very same set of constraints needed to obtain work at all.

But in addition, such systems achieve a "work task" closure. Call the three non-equilibrium processes in Figure 3.5 three *work tasks* to be performed. All three tasks are performed in a cycle. This is a work task *closure*.

The cycle of work here need not be a *thermodynamic* work cycle. This is because all three work tasks might be exergonic. But a work cycle linking both exergonic and endergonic work tasks might be involved, in which case a thermodynamic work cycle could be accomplished.

Further, each step is thermodynamic work. So not only is a task closure achieved, but also a cycle of work is achieved. Some machines, such as reciprocating engines, do a cycle of work. Not all do. If I lift a weight with a lever and fulcrum, a simple machine, no cycle of work has been accomplished.

Potential for Self-Reproduction:
Three Closures

I think it will be clear now that a system that exhibits constraint and work task closure may also potentially reproduce itself. The system is able to build the constraints that couple to non-equilibrium processes to do work to construct the self-same set of constraints. And in the process, it does a cycle of work.

All this happens in living cells, and these ideas will be central to the ideas we explore later about the origin of molecular reproduction. The spontaneous emergence of collectively auto-catalytic sets of polymers such as RNA or peptides is the case I will explain in subsequent chapters. Such systems achieve three closures: constraint closure; work task closure; and, as we soon shall see, something called catalytic task closure. In the latter, all the catalysts needed to make the same set of catalysts are contained in the autocatalytic set itself. These are three closures for life. The system literally constructs itself. We are on the verge of the organization of life.

Nonmystical Holism

The three closures—constraint, work task, and catalytic—are "holistic" properties, not to be found in any one of the parts alone. The three constraints, C_k, C_L, and C_i, mutually construct one another by the work task closures of processes 1, 2, and 3. Remove any one of these and the closures are lost. This "holism" is not mysterious but essential. Cells are wholes.

The Genericity of Physics and Specificity of Biology

Giuseppe Longo and Montévil (Longo and Montévil 2014, Montévil and Mossio 2015), write of the genericity of physics and the strange specificity of biology. "Mass" is a generic concept, they say. The mass of a teacup and of a rock may be the same.

With respect to falling, both are equivalent. The constituted objects of physics are generic, Longo notes: mass, position, momentum, and the symmetries of the laws of motion. In biology, sea cucumbers are sea cucumbers, however quickly they may fall from the Tower of Pisa. Rabbits are not sea cucumbers, although both may fall equally from Galileo's hand.

In physics, we need boundary conditions, but we tend to ignore where they come from. Part of the specificity of biology is that cells and organisms synthesize their own specific boundary conditions in Montévil and Mossio's constraint closure. The boundary conditions of cells are part of what we must study. Thanks to these, a mother rabbit builds a baby rabbit, not a tree.

Propagating Organization of Process

There are more things we need if we hope to explain the world outside our windows. Some of this we've seen: constraint closure and work task closure. And we've hinted at another: catalytic closure, to be explored in chapter 4. Then, as we shall see, comes enclosure into an "individual" in something like hollow lipid vesicle liposomes. This can yield a protocell, capable of heritable variation and selection. With all that, the total system can propagate its organization to yield a diversifying biosphere. Such systems literally construct themselves thanks to the three closures. As we will see, they evolve to create biospheres no one can prestate, governed by no entailing law at all. This life is thus reinhabited by a renewed and fully natural and nonmystical form of vitalism. Here, as Heraclitus said, the living world truly bubbles forth.

Chapter 4

Demystifying Life

The enormous problem of the origin of life, standing with the nature of consciousness and the origin of the universe as one of the three deep mysteries, was not even a problem before Pasteur. It was well known then that life arose spontaneously: maggots on rotten wood were abundant after a hard rain. What could be more obvious? Life arises freely.

Pasteur won a prize for a brilliant experiment. It was known that a sterile beaker of broth left open to the air soon became overgrown with bacteria. He crafted a beaker with an S-shaped neck filled with water, which acted as a stopper preventing bacteria in the air from reaching the sterile broth in the bottom of the flask. And now it remained sterile.

Life comes from life, declared Pasteur.

But if so, whence life in the first place? The origin of life problem was born—and remained almost mute for some 50 years until Alexander Oparin, a Soviet biochemist, suggested that life began as a coacervate, a gummy droplet, and J. B. S. Haldane proposed that the early oceans were a primitive soup of small organic molecules. Talks about the origin of life often show a Campbell's Soup can filled with a primitive broth.

The next major step was taken in the 1950s by young Stanley Miller, who set up a beaker with small organic molecules dissolved in water, with electrical sparks used to simulate lightning, and waited. A film of new molecules, rich in several amino acids, formed in the flask. Miller had demonstrated the abiotic genesis of amino acids—a hint that life could arise not just from life but from non-life. Over the ensuing decades, abundant work showed the abiogenic origin of sugars, amino acids, and nucleosides—the building blocks of proteins, DNA, and RNA.

Not long after, it was shown that the infall of meteorites on the early earth could have showered the planet with abundant organic molecular diversity. For example, the Murchison meteorite, landing near Murchison, Australia, in 1969, contains at least 14,000 organic molecular species. Thus a soup of organic molecules must have arrived from space. How thick the primitive soup may have been is unknown, but most workers look to abiogenic synthesis of organic matter on earth and meteoric infall during the formation of the earth as the two major sources of simple and complex organic molecules, the stuff of life.

The next issue, still unsolved, was the origin of molecular reproduction. We may know where the molecules came from, but how do they make more of themselves? Current cells have their complement of DNA, of RNA, of proteins, and of thousands of kinds of molecules coupled in a catalyzed metabolism, as well as myriad structures from lipids forming "bilipid" layers in cell membranes and organelles, to water inside cells. The cell as a whole reproduces. How did this arise?

The RNA World

The most obvious hypothesis for life's origin, suggested in the late 1960s, relies on the magnificent structure of DNA and RNA molecules. These form the famous double helix. As Watson and Crick noted with famous understatement in their 1953 paper on DNA, "It has not escaped our attention that the structure of the molecule suggests its means of replication."

Indeed, in DNA, the four bases—A, T, C, and G—show the well-known Watson-Crick base pairing, with A paired with T and C paired with G. Thus if, along one strand, or ladder side, of the double helix, there is the nucleotide sequence AACGGT, it is complementarily matched on the other ladder side by TTGCCA. The sequence of nucleotides along each strand specifies the nucleotide sequence along the other strand.

The same is true for RNA, which can also form a double strand helix. In RNA, U substitutes for T. This led Leslie Orgel, a chemist, to ask if a single strand of RNA, say CCGGAAAA, in a test tube could line up free G,G,C,C,U,U,U,U nucleotides, then, without an enzyme, link the latter into GGCUUUU to form a new complementary strand of RNA. At that point, the strand CCGGAAAA would be bound to the complementary strand GGCCUUUU, so the two would have to melt apart into separate single strands to replicate. Thereafter, each of the two single-strand systems would match up with more nucleic acids—G with C, A with U, and so on to create more double helices, which in turn would melt apart and start the cycle anew: a self-replicating system.

That's the theory anyway. The experiment is simple, brilliant, and should have worked. But it never has. This is partly true for good chemical reasons. The bond between two nucleotides in DNA or RNA is a 3'–5' bond, and this is less favorable thermodynamically than a 2'–5' bond. Here the numbers refer to atoms at different positions around the nucleotide. The 2'–5' bond does not allow helix formation. CCCCCCCCC can make GGGGGGGGG, but the latter single strand folds up and precipitates in the test tube precluding formation of a double helix. Others have tried molecules like DNA, called PNA. Nobody has gotten this to work for some 50 years or more. It may still happen. But research in this direction may be slowing down.

A major discovery, however, led in a different direction. In cells, protein enzymes carry out catalysis, speeding the molecular reactions crucial to life. It had been thought that only proteins catalyze reactions but that DNA genes and RNA were needed for proteins to be made. But about 20 years ago, it was discovered that single-stranded RNA molecules (called ribozymes) could also catalyze reactions.

Biologists were thrilled. The same class of molecule, RNA, could carry genetic information and catalyze reactions. Perhaps life was fundamentally based on a single class of polymers, RNA, which formed the scaffolding for what came after. This is the RNA World Hypothesis.

The central hope has been that an RNA ribozyme molecule might be able to copy itself! An RNA molecule is a sequence of nucleotides: A, U, C, and G. The idea is that the ribozyme could act on itself, nucleotide by nucleotide, in what is called "template replication," to self-replicate.

But how could one hope to find such a ribozyme?

A brilliant area of molecular biology now perhaps 25 years old is driving the search. It concerns starting with a known ribozyme and making a "soup" of that molecule and millions of different variants that are nearly but not quite the same. These are subjected to rounds of in vitro selection, then mutation experiments. For example, one can select for molecules that bind to a ligand or can catalyze the template reproduction of some target RNA molecule. The area is roughly known as "combinatorial chemistry." Using this, workers have evolved ribozymes in vitro and asked whether one of these creations could act as a ribozyme polymerase, an enzyme that could template-copy itself.

One such molecule has been found, which is able to copy a small part of itself. It was found starting with a ribozyme, gently mutating it to form a population of diverse molecules, and then selecting in vitro those showing some slight sign of a capacity to self-replicate. Then that selected population was mutated again and subjected to further selection over several cycles. The work is early but making real progress.

This would be a stunning success. As I note shortly, the RNA World is not my own view of the origin of molecular reproduction, but this is wonderful progress.

I think it is plausible that we will find an RNA molecule able to act as a polymerase to copy itself, and perhaps copy other RNA molecules as well. One rightly says to these workers, onward!

But I am also skeptical. First, how rare are such molecules outside of these difficult, deliberate experiments? If they are as sparse in nature as I fear—one molecule out of trillions of RNA sequences—then how would they have emerged to spark the creation of life?

Second, getting long RNA single-stranded polymers under plausible prebiotic chemical conditions is an unsolved problem in the first place. Third, would such an RNA sequence be stable in evolution, or would it melt away, driven by its own mutations of itself as it copied itself? The issue here is a problem called Eigen–Schuster error catastrophe. Manfred Eigen and Peter Schuster showed years ago that as the rate of mutation increased for an RNA sequence under selection, say in a test tube, at first the population stayed very close to the "master sequence." But as the mutation rate crossed a crisp threshold, the population veered away and became ever more different. The information in the master sequence is then lost. This is the error catastrophe. In short, under mutation and reproduction, would the RNA polymerase be stably reproduced hewing close to the master sequence?

But the problem is even worse. The Eigen–Schuster error catastrophe is for a fixed mutation rate. But what about our RNA polymerase? As it copies itself, the mutation rate should increase: the original master sequence RNA polymerase will be slightly likely to make mistakes on copying itself, introducing mutations into its daughters. Its slightly mutant daughter polymerases will be more error prone, that is, more likely to introduce mutations, so they will produce an even more wide-spread mutant-laden population of granddaughter offspring sequences, yet more error prone than their parents or the master sequence with which this all began. So the mutation rate is *increasing* with each round of reproduction. Thus the entire population of sequences should undergo an error catastrophe and diverge rapidly from where it began. This is easily studied. I gather from E. Szathmary (personal communication, September 2017) that this catastrophe occurs in theoretical studies. If this is correct, the nude replicating gene will melt away under the drive of its own error-prone reproduction. It is not evolutionarily stable.

Fourth, and perhaps the most important, the RNA polymerase able to reproduce itself is a *nude replicating gene*. It is merely an RNA sequence, floating defenseless. But how does this nude gene gather around itself a connected, catalyzed metabolism and a mechanism to synthesize lipids to form a liposome to house the RNA molecule and form a protocell? There is no obvious pathway from the ribozyme polymerase to these other feats.

I do not think my criticisms are fatal, but surely they are concerns.

The Lipid World

A second important strand in work on the origin of life begins with a different kind of molecule: lipids. These are long-chain fatty acid molecules with hydrophobic, water-hating, and hydrophilic, water-loving, ends. In a water environment, these form structures such as liposomes. The latter are hollow layer "bubbles" made from two layers of lipids, much like the membranes of a cell. The hydrophobic face of each layer opposes the hydrophobic face of the other layer, while the two hydrophilic layers are exposed to the watery environment outside and inside the liposome.

Liposomes are thus hollow lipid vesicles. Amazingly, David Deamer has shown that lipids from the Murchison meteorite can form these liposomes, suggesting how abundant in the universe life's building blocks may be. Additionally, if subjected to wet-dry cycles, for example, on surfaces of weather-exposed sand, liposomes can take in DNA and other polymers across their bilayer boundary. We'll discuss more about that later on.

Liposomes accomplish a simple miracle: they enclose an interior region of water and sequester it from the outside world. And in doing so, they prevent any molecular species trapped within that internal region from diffusing away, as would happen in an open, aqueous medium. Most of those interested in the origin of life are deeply fond of the idea that, however molecular reproduction may have begun, housing such a system in liposomes is a good idea. I shall be referring in the following to lovely ideas from David Deamer and Bruce Damer, 2015, on this subject.

But meanwhile, liposomes can grow and bud, forming two liposomes, hence achieving self-reproduction. This work was done by Luigi Luisi and also by Deamer. The capacity of liposomes to reproduce is at the heart of the lipid world view.

Doron Lancet has investigated what he calls the GARD model, for Graded Autocatalysis Replication Domain (Segre, Ben-Eli, and Lancet 2001). Here lipid molecules mutually catalyze the formation of one another in a collectively autocatalytic set and simultaneously form a spherical mass. Some numerical evidence suggests that the model can evolve the ratios of its molecular composition. Evolution in GARD is a step forward in the lipid world view.

Here too in the lipid world there are problems: it is not clear how to get from here to the other major classes of polymers, DNA and RNA, and peptides and proteins. The lipid world shows how to get the container but not the contents. And so we must move on, turning now to a body of theory about the emergence of molecular reproduction on which considerable work has been done by myself and others since I first introduced it in 1971 (Kauffman 1971, 1986, 1993; Hordijk and Steel 2004, 2017; Serra and Villani 2017; Vasas et al. 2012).

The Connectivity of Random Graphs

The first step in introducing these ideas is based on the work of Paul Erdős and Alfréd Rényi (1959) on the evolution of what they call random graphs.

A graph is simply a mathematical object with a set of dots connected by lines, or, put in more formal terms, vertices, V, connected by edges, E. A random graph is a set of vertices connected at random by a set of edges.

Erdős and Rényi asked what happens to a random graph as the ratio of edges to vertices, E/V, increases: that is, as more and more lines connect the dots. Figure 4.1 shows what happens. The result is remarkable. For E/V less than 0.5, the graph has a large number of disconnected "components." But when E/V crosses

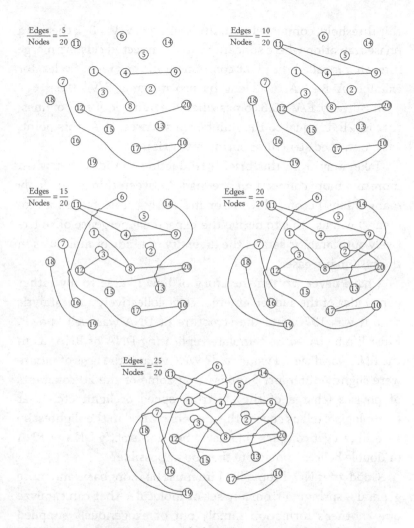

FIGURE 4.1 The Button Thread Phase Transition. Erdős and Rényi studied the evolution of "random graphs" where N nodes are connected by E edges, as the ratio of Edge to Nodes, E/N, increases. As E/N increases from 0.0 to 1.0 and higher, a "first order phase transition" occurs when E/N = 0.5. Before this, small clusters of connected nodes grow in size. At E/N = 0.5, suddenly a large connected cluster, or "giant component," forms with a diversity of "cycles." As E/N increases further, remaining isolated nodes become tied into the giant component.

this threshold, connected structures emerge. $E/V = 0.5$, then, is a phase transition where suddenly small connected clusters merge into what is called the giant component of the graph. Cycles, for example, A–B–C–A, of all lengths also appear at $E/V = 0.5$.

Intuitively, $E/V = 0.5$ comes when the number of *ends* of lines, that is, $2E$, is equal to the number of vertices, V. At this point, vast connected structures spring into existence.

Take away from this brief introduction the idea that when more and more connections are made between things, suddenly many things become directly or indirectly connected. In a moment, I will use this to derive the expected emergence of collectively autocatalytic sets as the diversity of kinds of molecules in the system increases.

I have never written the story of how I came to my rather happy idea of the sudden emergence of collectively autocatalytic sets. It was 1970, and the structure of DNA was well known. Must life be based on template-replicating DNA or RNA, as in the RNA world view, I wondered? Well, what if the laws of nature were slightly different? Suppose that some of the 29 constants of physics (charge of the electron, speed of light, etc.) that cosmologists tell us govern the universe varied in the slightest so we still can get complex chemistry but not exactly DNA or RNA or double helices. Would life then be impossible?

Good grief NO, I thought. Life must be more basic and more general—springing from *any* set of molecules that can catalyze one another's formation rapidly out of exogenously supplied building blocks. So what the universe needed was atoms, molecules, reactions, catalysis, and, well, something else. . . .

From this came the binary polymer model: at its heart are simple sequences, say peptides or RNA. ABBABBA is an example. Then we define simple reactions that our abstract polymers are able to undergo—ligation and cleavage such as AB + BAB = ABBAB or ABBAB = AB + BAB. Here, as the length,

N, of the longest polymer in the system is increased, the ratio of reactions per polymer, R/M, also increases, where R is the number of reactions and M the number of molecules. So the density of reactions per polymer would get higher and higher, pregnant with opportunity.

Now suppose that we speed up and intensify the reaction process by adding catalysts. And suppose that the catalysts consist of the very same polymers catalyzing the same set of reactions transforming polymers into one another. Maybe collectively autocatalytic sets would emerge!

Testing this idea experimentally with lab gear was not possible in 1970. So first, I made a simple model by assuming any polymer had the same fixed probability, P, of catalyzing any reaction. Later I improved this simple hypothesis. But the results are robust for either case.

Obviously, the idea was going to work. Think back to how a random graph of dots and lines, vertices and edges, undergoes a phase transition. As N, the length of the longest polymer, goes up, the *ratio* of reactions to polymers, R/M, goes up. If each polymer has a chance, P, to catalyze each reaction, at some point there are so many catalyzed reactions per polymer that about one reaction per polymer will by chance happen to become catalyzed and some analogue of Erdős–Rényi's giant component will emerge.

That is it.

It worked (Kauffman 1971). I was pretty thrilled when I simulated it. Then a week later, a famous theoretical chemist asked me why I was wasting my time with such nonsense, so I quit for a decade. Then, in 1983, at a conference in India on The Living State, I read the fine book by Freeman Dyson, *The Origins of Life* (1999), which proposed ideas similar to mine in 1971. I went back to work, and later (Kauffman 1986), and with Doyne Farmer and Norman Packard in 1986 (Farmer, Kauffman, and Packard 1986), published detailed simulations.

This work showed, and later work elaborates, that molecular reproducing collectively autocatalytic sets emerge as a phase transition in sufficiently diverse chemical soups of polymers that might be peptides or RNA or both (Farmer et al. 1986).

Such a set has intriguing properties. First, it exhibits holism. No molecule catalyzes its own formation. The set as a whole mutually catalyzes the formation of all the members of the set. The property is not present in any single molecule but distributed throughout the set.

Second, if we call catalyzing a reaction in the set a catalytic *task*, the system achieves *catalytic task closure*. All the reactions that need to be catalyzed are catalyzed. (I will join this closure to the other necessary ingredients—constraint closure and work task closure—shortly.)

Third, such a set is, like living organisms, a non-equilibrium system. Rather than succumbing immediately to entropy, it is fed by food molecules supplied from the outside. Thus, the non-equilibrium system can sustain itself by molecular reproduction. This is sounding more and more like what we call life.

Figure 4.2 shows a collectively autocatalytic set from Farmer et al. (1986).

In work done at the time (Kauffman 1993), it was shown that an improved model of which polymer catalyzes which reaction beyond the simple P previously mentioned also yields collectively autocatalytic sets. Here, any polymer needs to match its two substrates: for example, AAABAB would match BBBxxx on one end of one substrate, and xxxABA on the end of the other substrate, and only in the presence of such a match did the polymer then have a probabilistic chance to catalyze the ligation of xxxBBB to ABAxxx.

Work over the past 50 years shows that the model is robust: vary many of the details, and autocatalytic sets still easily emerge (Hordijk and Steel 2004, 2017).

THE ORIGINS OF LIFE: A NEW VIEW

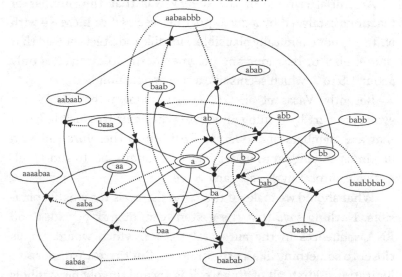

FIGURE 4.2 Collectively autocatalytic set. Circled binary strings are molecules, dots are reactions, solid lines lead from substrates to reaction dots to products. Dotted lines lead from molecules to the reactions they catalyze. Double circles are around the "food set" molecules supplied exogenously. Thus, the non-equilibrium system achieves constraint closure and work cycle. Function of polymers is to catalyze "next" reaction.

Work by Hordijk and Steel show that collectively auto-catalytic sets, slightly generalized to RAFs (reflexively autocatalytic and food-generated sets) that also allow rare spontaneous reactions, each consist in a number of irreducible RAFS, which combine into more complex RAFs. Each irreducible RAF consists of an autocatalytic loop and a tail of molecules that are catalyzed into existence but themselves play no role in autocatalysis. The entire RAF set consists of one or many different irreducible, autocatalytic sets together with one another.

An initial critique of the theory was that the number of reactions catalyzed by a given polymer seemed to increase with N. This is not chemically plausible. Hordijk and Steel showed that the number of reactions any polymer needs to catalyze is only about 1.5 to 2, which seems like a reasonable number.

Recently, Vasas et al. (2012) have shown that RAFS can evolve, in part by gaining and losing different irreducible RAFS that are members of the larger RAF, which therefore function like independent genes under selective conditions. In short, collectively autocatalytic sets can evolve.

What should we make of this? The theory is reasonably plausible. Nothing says that the system cannot mix peptides and RNA sequences in the autocatalytic set, which would get us closer to something like a primitive cell. But there are important limitations. First, all of this work is formal, involving symbols and algorithms rather than experiments in a dish. The appearance of a collectively autocatalytic set in a reaction graph is one thing. A real chemistry example may fail to reproduce. Serra and Villani (2017) have emphasized that the concentrations of constituents may be too low to be effective. However, in Farmer et al. (1986), we simulated this and found that reproducing sets emerged quite reliably. This warrants additional work. Further, Hordijk (personal communication, September 2017), has used what is called the Gillespie algorithm to study simple examples and finds reproduction emerges quite reliably. The Gillespie algorithm allows one to study chemical systems in which the number of copies of each molecular species is very small. In the RAFs that Hordijk and Steel study (Hordijk and Steel 2004, 2017), uncatalyzed reactions can happen slowly and spontaneously, so the members of the set need not be there from the start. By such spontaneous reactions, an RAF can come into existence without having to have had all its components present initially. All this is very encouraging, but, as usual, much work needs to be done.

A second limitation is that a connection to the lipid world and a way to enclose the molecules into a cell-like sac is so far lacking. The ideas of Damer and Deamer (2015), described in a later chapter, may bridge this gap.

From the Computer to the Lab

Collectively autocatalytic sets have been created of DNA, of RNA, and of peptides. I'll explore these one by one.

DNA Collectively Autocatalytic Sets

In the mid-1980s, the first molecular reproducing system was made by G. von Kiedrowski (1986) using real DNA—a sequence of six nucleic acids: CGCGCG. Von Kiedrowski crafted this hexamer, as it is called, and two short trimers that complement two halves of the molecule: that is, GCG, which complements the CGC on the "left," and CGC, which complements the GCG on the "right." In solution, the hexamer binds the two trimers by Watson–Crick base pairing, and catalyzes the ligation of the two trimers to make a new hexamer, GCGCGC. Then, read from right to left, the new hexamer is identical to the initial hexamer. Thus this little system reproduces itself.

The reaction is autocatalytic. Moreover, the hexamer serves as a simple "ligase," ligating the two trimers. The hexamer does not, however, act as a polymerase, adding the nucleotides one by one in template replication. Thus, here, molecular reproduction of DNA occurs without template replication, as imagined in the RNA World.

Shortly thereafter, von Kiedrowski created the world's first collectively autocatalytic set consisting of two different hexamers, each of which could reproduce the other.

The conclusion is that collectively, autocatalytic sets of small polymers can be constructed and work.

Peptide Collectively Autocatalytic Sets

Proteins have been thought incapable of reproduction, for they have no axis of symmetry like the self-complementary DNA helix. But this firm thought turned out to be wrong. In 1995, R. Ghadiri made a self-reproducing small protein! He started with a protein that forms a helix that coils back on itself. Ghadiri reasoned that one part of the coil could bind to and recognize the other part of the coil, so he took two shorter fragments of a 32-amino-acid-long region of the protein and incubated them all together. The longer sequence bound the two shorter ones; then further, the longer sequence catalyzed the formation of a peptide bond between the two shorter fragments, forming a second copy of the original sequence. The system reproduces itself.

Even proteins can do it!

A few years later, G. Ashkenasy, a postdoctoral fellow with Ghadiri, had formed a collectively autocatalytic set consisting of nine peptides. I will describe this in detail in the following.

The conclusion is that molecular reproduction by systems of small proteins is clearly possible. Therefore, molecular reproduction need not be based on the template replicating properties of DNA, RNA, or similar molecules. Moreover, the prebiotic synthesis of amino acids is quite easy, as is the formation of small peptides; thus, the possibility of the early and spontaneous emergence of collectively autocatalytic sets of peptides must be taken seriously.

RNA Collectively Autocatalytic Sets

Recently, two bodies of work have achieved RNA collectively autocatalytic sets. Lincoln and Joyce have used in vitro selection, as described previously, to evolve a pair of ribozymes that catalyze one another's formation by ligating two fragments of each (Lincoln and Joyce 2009).

Lehman and co-workers (Vaidya et al. 2012) carried out a stunning experiment using a set of ribozymes, each cut in half to isolate the recognition site and the catalytic site. The recognition site recognizes the target of the ribozyme; the catalytic site carries out the catalytic task of the ribozyme. These halved ribozymes were then incubated together. The catalytic site of one could bind to the recognition side of another to create a functioning hybrid ribozyme. The system formed a single autocatalytic ribozyme, then loops of 3-, 5-, and 7-member collectively autocatalytic sets! The multimembered RNA sets outcompeted the single autocatalyst. This is the spontaneous emergence of self-reproduction from a pool of molecules, a remarkable find.

So RNA molecules too can form collectively autocatalytic sets.

Lehman's wonderful result still starts with highly evolved RNA ribozyme sequences. The aim is to achieve the spontaneous formation of collectively autocatalytic sets from a pool of unevolved RNA; for example, random RNA sequences, or other molecules, such as random peptides, or both. Experimental work is now pursuing this. Keeping in mind Serra's caveats about low concentrations limiting the emergence of such sets, we can hope that experiments will soon demonstrate the spontaneous formation of autocatalytic sets from unevolved RNA, peptide, or other molecular sequences.

Three Closures of Life

In chapter 3, we discussed the remarkable concept of Constraint Closure due to Montévil and Mossio (2015). And we introduced a second idea: work task closure. The former is the fact that a set of constraints on non-equilibrium processes can do work to construct the same set of constraints. The latter is the set of thermodynamic work tasks that are done to accomplish this. We now show that collectively autocatalytic sets achieve these and a further, third closure: catalytic task closure. Such systems are inherently open, non-equilibrium systems and reproduce.

Gonen Ashkenasy has a nine-peptide collectively autocatalytic set flourishing at the Ben Gurion University in Israel (Wagner and Ashkenasy 2009), Here, peptide 1 catalyzes a reaction forming a second copy of peptide 2 by joining two fragments of peptide 2. Peptide 2 catalyzes the formation of a second copy of peptide 3. Peptide 3 does the same for peptide 4, then 5, 6, 7, 8, and 9, with peptide 9 catalyzing the formation of a second copy of peptide 1 to close the catalytic cycle.

The three closures are achieved. First, there is catalytic task closure. No peptide catalyzes its own formation. Each of the nine reactions requiring catalysis is catalyzed by one of the nine peptides. Work task closure is realized. Each of the reactions is a task, and work is done performing that task, seen in the formation of a new peptide bond in the ligated product. So a cycle of real thermodynamic work is carried out. But in addition, each peptide *as the catalyst* is a boundary condition constraint on the release of energy. The catalyst binds the two substrate fragments, holding them close together and lowering the energy potential barrier of that ligation reaction. This is precisely a constraint altering the release of energy into a few degrees of freedom. The catalyst is like the cannon, constraining the release of energy to fire the cannon ball. So the nine peptides *are* the nine constraints, and this system achieves precisely Montévil and Mossio's constraint closure. The peptides, as the catalysts, are the constraints; and the system builds second copies of its own constraints on the release of energy in each of the nine reactions.

Finally, the system is displaced from equilibrium by the continued feeding of the two fragments of each of the nine peptides. Ashkenasy's set realizes the three non-local closures—constraint, work, and catalytic—in a far from equilibrium system. These are the hallmarks of life itself. Cells do the same things.

Life as the Daughter of Molecular Diversity

An RNA molecule able to act as a polymerase and copy itself is no less or no more than a nude replicating gene. Life in this view started simple. Molecular diversity is minimal. A single sequence would do it. It may work, and God speed.

But the Murchison meteorite has at least 14,000 diverse organic compounds. The early earth may have had similar molecular diversity from very early on by meteor infall and from in situ synthesis. Thus, molecular diversity was almost surely at hand.

The theory of the spontaneous emergence of collectively autocatalytic sets explicitly builds on molecular diversity. Such sets emerge as the diversity of the chemical soup crosses a critical diversity threshold where connected catalyzed webs of reactions among the constituents arise. The holisms of catalytic, work task, and constraint closures are daughters of this diversity.

My strong suspicion is that life emerged not nude and simple but whole and complex, as a web of reactions that mutually catalyzed one another. We will confront in chapter 5 the issue of the formation of a catalyzed reaction network among small organic molecules that constitutes an early metabolism. High diversity and a multiplicity of potential polymers that can act as catalysts may help the formation of a metabolism. No one knows, but that is where I'm placing my bet.

Life arose in about 200 million years. The routes to life must be reasonably probable, not hidden in some recess. Perhaps life is the daughter of molecular diversity.

The "Life Force"

Until a little more than a century ago, many scientists believed in a mysterious "life force"—elan vital, vitalism. With the synthesis of urea, it was realized that biological organic molecules were ordinary chemicals, so life need not be based on some mystical phenomenon.

With the three closures, we have a holism that need not invoke any ineffable magic. In a collectively autocatalytic set, the three closures are not properties of any single molecule but of an interwoven set of molecules and reactions. I suspect that together, these three closures constitute "elan vital," a nonmysterious but wonderful life force. By the constrained release of energy into a few degrees of freedom, such non-equilibrium systems can do real thermodynamic work and can construct and reproduce themselves.

We will not find any of these three closures in any single molecule or reaction in the system. These are properties of "wholes"; but again there is no mystery, no new force, rather a new organization of matter, energy, entropy, constraint, and thermodynamic work into a whole that is, I suspect, the center of life itself.

Life is a fundamentally new linking of non-equilibrium processes and boundary condition constraints on the release of energy into a few degrees of freedom that thus is thermodynamic work. But stunningly, the work done can *construct* constraints on the release of energy in further non-equilibrium processes. In reproducing systems such as cells, a closure is achieved linking these processes and constraint construction into an organization that closes on itself. The system, in doing this work, constructs its own constraints and also reproduces, achieving catalytic task closure.

Such a system is a "machine" but not of matter alone, energy alone, free energy alone, entropy alone, or boundary conditions alone. It is a new union of these.

Cells do cycles of work to construct approximate second copies of themselves as physical objects when they reproduce. Trees do cycles of work to construct themselves when they grow from seeds. These are examples of propagating work and the propagating organization of process in the living world. The evolving biosphere *is* this co-constructing propagation, subject to heritable variation and natural selection. This is how the evolving biosphere physically builds itself and evolves. It surges upward in complexity and diversity into the indefinite, nonergodic universe above the level of atoms. Hearts come to exist.

We may have found the "life force": not a nonphysical mystery but a marvel and a different mystery of unprestatable becoming.

Chapter 5

How to Make a Metabolism

It is a torpid afternoon 3,786,394,310 years ago in a hot spring on what later would be Western Australia. More precisely, it is 3:17 p.m. local time.

A nude replicating RNA ribozyme polymerase, James, is, well, reproducing. "And one and two and one and two," mumbles James, adding nucleotides one after another to a copy of himself. "Whew, that was a heave," thinks James, who lets go of the newly synthesized copy of himself.

"Now where was I? Oh yes, get on with it . . . And one and two and one and two," he repeats making another copy of himself. . . .

"In the end, kind of boring," he sighs.

"If only I had a metabolism, all rich and varied with . . . Oh well, I don't."

James tries to imagine how he can get himself surrounded by and ensconced within a rich metabolism that would make his work so much easier. If he had a metabolism, he could synthesize his own nucleotides and not have to wait for them to diffuse to him in the dilute hot spring. But he cannot imagine how.

How indeed does this arise?

There is nothing wrong with nude replicating genes, bless them, but the next big leap—getting to a connected web of catalyzed chemical reactions that supports the nude gene and is somehow supported by it—seems rather a puzzlement.

I want to suggest that life did not happen this way, with a nude gene in a near sterile environment and with no metabolism. The Murchison meteorite has, as noted, a diversity of 14,000 organic molecules in a smallish chunk of that which came to us from afar, from when the solar system was forming. If the meteorite could be chemically diverse, then by infall of such material to the early earth, and in situ synthesis, the spring was presumably diverse chemically as well.

Thus the hot spring was undoubtedly chemically highly diverse. Can we make use of that to think about the origin of both autocatalytic sets, as in chapter 4, and a connected catalyzed complex metabolism? And might the metabolism help the autocatalytic set and vice versa? Yes.

I shall claim that our metabolism, like the rest of life, is the daughter of this diversity.

Figure 5.1 shows a graph of human metabolism. Dots are species of molecules, lines are reactions. A metabolism is a huge web of reactions among these molecules, almost all of which are catalyzed by specific protein enzymes encoded by genes.

The reactions of metabolism don't happen magically. Instead, each reaction requires that the reactants have more chemical energy than the products and that energy is used in the reaction. In other words, all metabolism is driven by an input of chemical energy at the top of the metabolic sequence, with products at the bottom containing less chemical energy. In the biosphere, the energy at the top is supplied by chlorophyll-capturing photons, then releasing high-energy electrons to

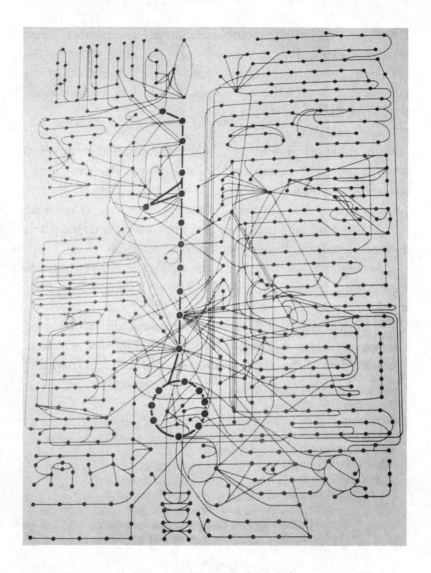

FIGURE 5.1 Human metabolism. Stuart Kauffman, The Origins of Order, Oxford UP, 1993.

NADP (nicotinamide adenine dinucleotide phosphate). Those electrons carry the chemical energy downhill through the metabolic chain. In the last step, they are stripped off in the citric acid cycle with the release of CO_2, which is the sink at the bottom of the chain.

How could this arise?

I am going to suggest that a catalyzed connected metabolism arose as exactly the same kind of Erdős–Rényi phase transition we saw in chapter 4, where collectively autocatalytic sets spontaneously emerge as the diversity of molecules crosses a threshold. Such sets reproduce, thus molecular reproduction emerged whole and connected, and I think metabolism did as well.

The hypothesis is a bit radical but fully testable. To begin to tune our intuition toward this idea, I present Figures 5.2a, 5.2b, and 5.2c.

In these figures, different dots represent different molecular species; boxes represent reactions. Black arrows lead from substrate dots to the corresponding reaction box and from that

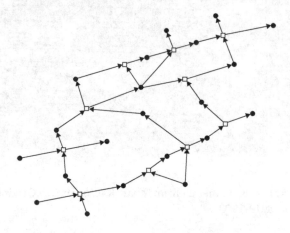

FIGURE 5.2A Chemical reaction graph: dots = molecules, boxes = reactions (e.g., small molecule reactions forming amino acids).

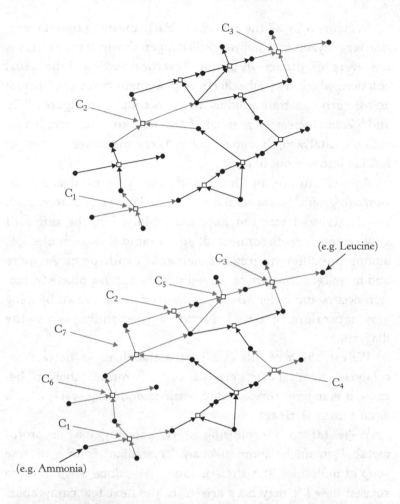

FIGURES 5.2B AND 5.2C Light arrows are catalyzed reactions. The dotted arrow from each catalyst, C_i, goes to the reaction it catalyzes. Each catalyst is also a constraint and boundary condition.

reaction box to the product dots from that reaction. This is called a "bipartite graph," because there are two elements, dots and boxes. Each dot connects only to boxes, and each box connects only to dots.

In Figure 5.2a, all the arrows are black, meaning that no reaction is catalyzed, but they may still happen slowly. If the reactions are reversible, the arrows do not show the direction of the actual reaction, which instead reflects displacement from equilibrium across each substrate product reaction couple. In Figures 5.2b and 5.2c, an increasing number of reactions are catalyzed. If a reaction is catalyzed, that box is colored grey and so are the arrows leading into and out of it.

As you can see, in Figure 5.2b, the "grey catalyzed reaction sub-graph" has several disconnected grey structures. Each small catalyzed reaction happens rapidly, but the different grey regions are disconnected, so no rapid flow of molecules among the different grey sub-networks can happen. As more and more reactions are catalyzed (Figure 5.2c), a phase transition occurs and order arises spontaneously, as shown by a big grey network of catalyzed reaction lines spanning across the diagram.

We can think of this as a protometabolism—a network of catalyzed chemical reactions. It is only a "protometabolism" because it is not yet connected to a self-reproducing system, such as an autocatalytic set.

We're off to a promising start, showing how a protometabolism might spontaneously "crystallize" from a diverse soup of molecules. But there is more to be done if we want to suggest how life may have arisen. In the next few paragraphs, I will take us through a few more steps.

First, we want our protometabolism to be connected to an autocatalytic set, so that the set's own molecules (peptides or RNA) act as the very catalysts that catalyze the connected reactions. And then we want the products of that metabolism to feed back into the autocatalytic set so they jointly "help one another." *If* this can arise, both the metabolism and the autocatalytic set can jointly support one another and co-evolve.

In Figures 5.2b and 5.2c, I show specific catalysts catalyzing each reaction. These, again, could be peptides or RNA drawn from a collectively autocatalytic set.

I give a toy example. Suppose, as in chapter 4, that we have a set, C, of candidate catalysts that are peptides, and each such peptide has a fixed probability, P, of catalyzing each of the reactions in the whole reaction graph shown in Figure 5.2a. Then the number of catalyzed reactions, Rc, depends on the number of reactions, R, in the system: here, R = 10, and on the value of P, say 10^{-2} or 1/100, which is the probability any candidate catalyst catalyzes any chosen reaction. And it depends on the number, C, of peptide candidate catalysts thrown at the system. Suppose C is, say, 100. Then the expected number of catalyzed reactions is Rc = RPC = 10. Thus, under these assumptions, all or almost all of the reactions in Figure 5.2a will be catalyzed. This is the same as the Erdős–Rényi transition. We expect all 10 reactions to be catalyzed, so the whole network is connected and a giant grey catalyzed reaction structure will arise, as in 5.2c. We have crossed the threshold of the Erdős–Rényi phase transition. And so a connected catalyzed reaction sub-graph forms.

The toy example shows us that if there is a reaction graph with some large number of reactions, R, and a much smaller number of molecules, N, then when N reactions are catalyzed, a giant connected, catalyzed, reaction graph will be formed. For any choice of the probability of catalysis, P, a correspondingly high enough number of candidate peptides or other catalysts ensures that the peptides will catalyze a connected metabolism.

CHNOPS

Until this point, we have been dealing with abstract molecules and abstract reactions. But will our ideas work in our own world? As we add real molecules to a lifeless broth, will we reach a point where the soup becomes so diverse that a phase transition occurs, with isolated clumps of chemicals springing together to form a self-sustaining, catalyzed, connected metabolism?

The organic molecules we know on earth are made from atoms of carbon, hydrogen, nitrogen, oxygen, phosphorus, and sulfur—CHNOPS, for short. So let's use these to form a reaction graph. We are still dealing with hypothetical cases, but this will take us another step toward the concrete. Again, molecular species (real ones!) are dots, reactions are boxes, and black arrows lead from substrates to boxes and from boxes to products.

Molecules are made from atoms—CHNOPS—so we must introduce another variable: M for the number of atoms per molecule.

Consider a reaction graph with up to M atoms of CHNOPS per molecule. For example, if M = 1, we just have lone atoms, C, H, N, O, P, and S. If M = 10, we have molecules consisting of up to 10 atoms chosen among C, H, N, O, P, and S.

It's easy to see that as M increases, say from 1 to 100, the number of molecular species, N, shoots up very rapidly. Organic molecules can be complex beasts bristling with side chains. But for now let's keep it simple. Consider the hypothetical case of linear polymers—single strings—made from just two building blocks, A and B. Here the number of molecules with up to M atoms is 2 to the M + 1.

Now we ask, as M increases, how many reactions, R, among the N molecular species are there? In general, R will increase even faster than N. And the ratio of reactions R to molecular species N, as a function of M, is (M − 2) (Kauffman 1993). In other words, the ratio of reactions to molecules is essentially M, the length of the longest polymer in the system. In short, the number of reactions per molecule goes up rapidly as molecular complexity increases. This is just what we want.

At this point we are dealing with uncatalyzed reactions. As M increases, the molecular network becomes ever denser with more reactions per molecule. It is this fact that helps drive the emergence of the Erdős–Rényi phase transition: when there are enough reactions in the system, many of them will, just by chance, be catalyzed by a sufficiently diverse set, C, of candidate catalysts. A grey sub-graph of connected catalyzed reactions arises connecting the molecular species in the system. Turning up C, or turning up M, drives the phase transition.

This is shown in Figure 5.3 as a hypothetical hyperbolic curve in a coordinate system with two axes. The x axis is the diversity, C, of candidate catalysts; the y axis is N, the number of molecular

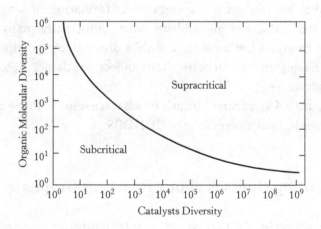

FIGURE 5.3 Photograph of a hyperbolic curve.

1. Substrates and products dots. N dots.
2. Reaction boxes with lines connecting dots, R reactions.
3. Bipartite graph.
4. CHNOPS = carbon, hydrogen, nitrogen, oxygen, phosphorus, and sulfur.
5. Consider all molecules, N, up to M atoms CHNOPS per molecule.
6. How does N increase with M, i.e., number of molecules with M atoms per molecule?
7. How many reactions, R, among the N?
8. Ratio of R/N. It should increase as N increases (as in binary polymer model).
9. Catalyze a fraction, F, of the R reactions.
10. Catalyzed reaction subgraph.
11. What is the size of largest connected component of the catalyzed graph as a function of F and N?
12. What is the descent distribution of catalyzed graph as a function of F and N?
13. Implications of descent distribution on matter transport in graph.
14. If a set of C molecules and peptides, and each catalyzes each reaction with probability P, what is the catalyzed graph as a function of P and M and C?

FIGURE 5.4 Chemical Reaction Graphs

species in the reaction graph. Below the curve, the system is "subcritical" and the Erdős–Rényi transition has not happened. Above the curved line, the system is supracritical, and the phase transition has occurred.

Voilà! We can get our spontaneous formation of a holistic connected catalyzed metabolism when sufficiently many candidate catalysts are incubated with a diverse enough chemical soup having more reactions than molecules. James need not struggle alone.

Figure 5.4 summarizes much of what we want to know about the chemical reaction graphs of CHNOPS.

Beginning Guesses about Reaction Graphs

We saw previously that given a sufficiently diverse soup of molecules, there is a chance that some will be of just the right shape to act as catalysts, resulting in a self-sustaining web of chemical reactions. But can we calculate the odds that this will occur?

Little is yet known about the structure of real chemical re-action graphs among CHNOPS as M increases—that is, the number of molecules and reaction networks that arise among them. For now, we will try to calculate a plausible number using our simple hypothetical organic molecules made of linear polymers of two kinds of atoms, A and B (see Table 5.1). Figure 5.5 shows a case where we have a metabolism consisting

---

TABLE 5.1 Critical Number of Hypothetical Organic Molecules Made Up of Linear Chains of Atoms A and B Required for Crystallization of a Connected Metabolism

$$\bar{P} = e^{-P(5000)(M-1)\left(1+2^{M+2}\right)} = \frac{1}{e^8} < 0.001$$

P	M	2^{M+1}
10^{-4}	1.965	8
10^{-5}	3.81	28
10^{-6}	6.25	152
10^{-7}	8.98	1,010
10^{-8}	11.85	7,383
10^{-9}	14.83	58,251

Note: Number of potential catalysts present is 5,000. Note that the number of hypothetical organic molecules would suffice for the emergence of a connected metabolism. Thus, in a two-dimensional space whose coordinates are the number of catalysts and number of organic molecules in the system, each value of *P* determines a critical curve separating regions with and without connected metabolisms.

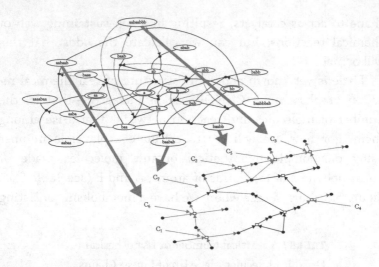

FIGURE 5.5 Coupling peptide autocatalytic set and metabolism (the peptides are the catalysts for the metabolism!).

of 5,000 candidate catalysts (C) and can tweak the number of atoms (M) and the probability (P) that a molecule will catalyze a reaction. Because the equation (derived in my book *Origins of Order*; Kauffman 1993) in Figure 5.5 is continuous and highly nonlinear, the results are given in real, not integer, values of M, thus approximating what will work for integer values of M. M, the atoms per molecule, is a stand-in for the true diversity, N, of a molecular system.

As you can see, the values of P range from 10^{-4} to 10^{-9}. In other words, the chance that a random peptide catalyzes a reaction ranges from one in 10,000 to one in a billion. For $P = 10^{-7}$ and C = 5,000, the connected metabolism has only about 1,000 molecular species, with length up to about 9 monomers of A and B.

Interpolating crudely, if the probability of weak catalysis by a peptide is 10^{-5}, only about C = 150 peptide candidate catalysts are needed to achieve a connected catalyzed metabolism of several dozen molecular species.

These are encouraging results. We want the collectively autocatalytic set to catalyze the reaction of the metabolism. Here such set with 150 members can do so, with a metabolism of several dozen small molecules.

This could yield a first step in which a collectively autocatalytic set catalyzes a small metabolism that resides in the same vicinity, "side by side" if you will. If the products of the metabolism can feed the autocatalytic set (see following), the two can co-evolve: a self-reproducing system and its supporting catalyzed connected metabolism.

These are very crude calculations, based on a crude model of the real chemical reaction graph for CHNOPS. The simple model here serves three purposes: First, it shows that the Erdős–Rényi phase transition is robust across a wide range of assumptions. Second, it suggests what the proper theoretical calculations for the real CHNOPS reaction graph might be showing with respect to the expected sizes of catalyzed reaction graphs as a function of C, the number of candidate catalysts, and P. And finally, it suggests how we can test the idea with experiments.

Into the Laboratory

Before we proceed, we must consider what has been learned from other experiments that bear on the origin of life issue. The emergence of collectively autocatalytic sets depends on the chance that a random polypeptide catalyzes a randomly chosen reaction. We know a bit about this.

What Is the Chance That a Random Peptide Can Fold?

Proteins are linear sequences of amino acids that then fold up to make mature proteins able to catalyze reactions and perform cellular functions. We begin to know the answer to how likely it is that a random protein sequence folds from work in my laboratory by Thom LaBean in 1994 and 2010 and Luigi Luisi in 2011. They showed that about 20 percent of random sequence polypeptides fold. The data is very crude but can easily be improved. If folding is necessary for function, it lies readily at hand.

What Is the Chance a Random Peptide Binds an Arbitrary Ligand?

Using phage display, it appears that the answer is about 10^{-6}. Phage display is a technique in which genes coding for random peptides are cloned into a viral phage coat protein gene, so the peptides are "displayed" on the surface of phage viruses. Then phages with different random peptides are tested to see if any of them bind to a given target molecule. Initial experiments by G. Smith showed that out of about 20,000,000 sequences, 19 different hexamer peptides were found to bind a given monoclonal ligand. This yields $P = 10^{-6}$.

The criteria used in these experiments for selection is "adequate" binding to the monoclonal ligand under experimental conditions. We do not yet know the probability of *weak* binding, but this can be explored. If 10^{-6} is the probability for reasonably strong ligand binding, then 10^{-5}, as I just hinted, is not a foolish value for weak binding, hence possible catalysis.

What Is the Probability a Random Peptide Catalyzes a Randomly Chosen Reaction?

Here we can make a good guess. Work some 20 years ago or more showed that molecules called monoclonal antibodies can catalyze reactions. (Monoclonal antibodies are sets of identical antibody molecules.) These were made by finding a monoclonal antibody that bound to a molecule that is stably shaped like the transition state of a reaction; hence, it is called a stable analogue of the transition state of a reaction. Such a monoclonal, with high probability, also catalyzed the reaction. Assuming that the probability of a monoclonal, like a phage display, binding a random molecular bump is 10^{-5} to 10^{-6}, including binding a stable analogue of a transition state, the probability of catalysis is also one in a hundred thousand to one in a million!

What all of this comes down to can be succinctly stated: Random polypeptides often fold and can with a probability of about one in a hundred thousand to one in a million bind a molecular epitope, or bump, and catalyze a given reaction.

We are ready now to consider how to test these ideas in laboratory experiments. To get a sense of how feasible our ideas are, first we want to further assess the probability that a random peptide catalyzes a random reaction. Consider that a reaction has a product that may be detectable even at very low concentrations. For example, the product might bind to a protein receptor and thereby alter its flow behavior, a phenomenon that can be seen at extremely low (10^{-15} molar) concentrations.

Suppose the probability that a peptide catalyzes a reaction is, we might think, 10^{-6}. Put 100 samples of 10^4 out of a "library" of a million different peptides in 100 reaction vessels. Incubate each vessel with the substrate of the reaction and test for the desired product. If that product is found, we can conclude that one or more peptides in that pot carries out the reaction. Now we narrow it down, dividing that pot into 100 pots each with 100 of the peptides and rerun the experiment. Then we find the pot or pots with the desired catalyst making the product and iterate again to 100 pots with 1 peptide per pot.

If this process yields C catalyst peptides that catalyze the reaction, then $P = C \times 10^{-6}$ is the rough probability of catalysis.

A modified version of this reaction uses a set of R independent reactions in which each yields a discriminable product. Here, we aim to show we can catalyze a set of R reactions using a set of random peptides as candidate catalysts. We will use this in a moment.

Now we can go a critical step further to the Erdős–Rényi phase transition: can we catalyze the formation of a connected catalyzed reaction graph among some of the many N by using a large enough set of candidate catalysts, C?

Consider a reaction set, with N organic molecules that belong to a complex reaction graph consisting of R reactions, and throw a set, C, of random peptides into the mix.

We know the Erdős–Rényi threshold will be crossed for any fixed probability of catalysis, P, by tuning C and R. And so we can vary C or R and test to see what the probability is that catalyzed reactions begin to occur. This can easily be tested experimentally. Suppose we start with members of CHNOPS up to M = 6 atoms per molecules. If reactions are catalyzed, forming larger molecules, for example, M = 10 or 15, from a founder set with maximum size M = 6 atoms, this is easy to see by mass spectroscopy or high pressure liquid chromatography. Both are highly sensitive techniques to test for the size of a molecule. Further, there is a critical curve of the phase transition in the C R plane where we expect to see catalysis set in. One can map this out by tuning C and R. So we can test for the onset of catalysis as R and C vary.

We can also test for the point where the Erdős–Rényi phase transition occurs, spontaneously forming a giant catalyzed component in the reaction graph. We can do this by looking at the transport of material across the catalyzed reaction graph.

If a giant connected catalyzed component is formed, then transport of matter should happen across the catalyzed graph! What would this look like? Can we posit a theory and test it experimentally? Yes. For example, imagine isotope labeling the nucleus of an atom in a small molecule. If the molecule is part of a connected pathway of catalyzed reactions, the isotope labeled nucleus should be transported, sometimes to the larger molecules, via that catalyzed pathway. This is directly testable by mass spectrophotometry. I find this very exciting. We will not necessarily know at first the detailed structure of the catalyzed reaction graph that we form, but we can test the flow of labeled nuclei down connected pathways from an initial labeled nucleus in an initial labeled molecule to all its descendants in the graph. From the data, we can *infer* the structure of the catalyzed reaction graph.

Clearly, it is a well formulated theory task to ask the structure of the graph by which molecules are connected to which other molecules by catalyzed reactions in a reaction graph subject to random (or nonrandom) catalysis. Furthermore, the flow of matter on such a graph is subject to theoretical investigation. For example, one can simulate the flow of matter, or nuclei, among atoms and molecules on this graph.

Indeed, we can discover the connectivity structure showing which molecules are connected by catalyzed reactions to which molecules of the giant component: define the descendants of a node in a graph as those reachable from that node along grey arrows. Define the radius from the node as the shortest path to the most distant descendant. Define the descent distribution and radius distribution over the graph. What do these properties look like as a function of C, P, and R? Using isotope-labeled nuclei, all this can be tested by the flow on the graph.

I suspect that such research will help us understand how metabolism arose billions of years ago on an early planet called home.

Joining a Metabolism to a Collectively Autocatalytic Set

We have now shown how a connected catalyzed metabolism can have arisen by the Erdős–Rényi phase transition. We need now to ask if this separate metabolism can be reciprocally connected to an autocatalytic set, the former feeding the latter, and the latter catalyzing the reactions in the metabolism.

In short, can we marry our catalyzed metabolism to a collectively autocatalytic set so the set catalyzes the reactions in the connected metabolism, and the metabolism makes molecules useful to the set?

Of course we can, for it is our hypothesis, and a perfectly reasonable one.

Figures 5.6 and 5.7 show our happy system. In Figure 5.6, some of the peptides in the autocatalytic set catalyze some of the reactions in the metabolism. In reality, of course, we hope that peptides will catalyze *all* the needed reactions in the metabolism. In Figure 5.7, the metabolism reciprocally feeds the collectively autocatalytic set! That is, the metabolism makes small molecules needed by the set. Thus, the metabolism and autocatalytic set are married.

Now we are ready for an important new idea. Collectively, autocatalytic sets are composed of a number of irreducible autocatalytic sets. They are irreducible because if you remove one type of molecule, the self-sustaining structure collapses. Each such irreducible set has an autocatalytic loop, and probably has a tail of one or more peptides hanging off the loop, but those play no

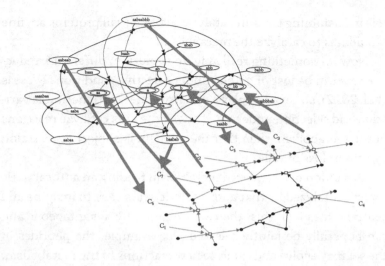

FIGURE 5.6 The metabolism "feeds" the autocatalytic set; the set catalyzes the metabolism.

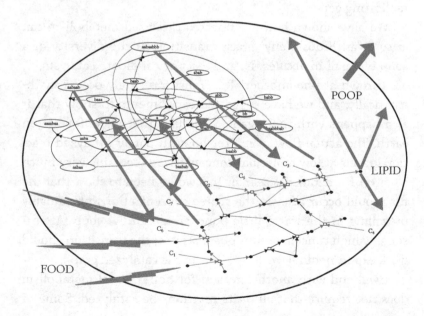

FIGURE 5.7 The protocell.

role in sustaining the autocatalytic set. Such tail proteins are fine candidates to catalyze the metabolism.

Now for something really big: in evolution, different irreducible sets can be lost or gained under selective conditions (Vassas et al. 2012), so they can play a role like genes, with heritable variation and selection underway. Thus, if the tails catalyze reactions in the metabolism, then that metabolism is subject to selection by gain or loss of tails.

This union of a catalyzed metabolism feeding an autocatalytic set and catalyzed by that autocatalytic set is easy to imagine and central—for given this, the two are mutually advantageous and can hopefully be jointly selected. For example, the peptides in the set may evolve and catalyze new reactions in the metabolism, generating new kinds of molecules to feed the set.

We have come a long way from James, the laconic nude replicating gene.

We have shown how a connected catalyzed metabolism can arise as an Erdős–Rényi phase transition as the diversity of a soup of small molecules, N, and a set of candidate catalysts, C, are thrown at one another. The transition surely occurs mathematically, and we have suggested experiments to show that it also happens with the organic molecules that make for life on earth. The actual flow of matter on such partly catalyzed reaction graphs at low molecular concentrations remains very much a matter for detailed research. This would need to show that the flow could occur at even the sparse concentrations presumably prevailing in the warm little pond of primordial soup. At reasonably high concentrations of substrates, the transition should work experimentally with real flow in the catalyzed graph.

I will end with another reason for hope. A real metabolism does not require that all of its reactions be catalyzed. Some of them can happen spontaneously. In *E. coli*, in fact, there are some 1,787 reactions, and only three are uncatalyzed (Sousa et al. 2015). Remarkably, *E. coli* metabolism is itself collectively autocatalytic (Sousa et al. 2015). Autocatalysis is present in living cells.

We are ready now for the protocell.

Chapter 6

Protocells

No one knows how life started. But many workers think that early life began with what they call "protocells." A protocell is imagined to be some kind of self-reproducing molecular system, perhaps coupled to a metabolism and housed inside a hollow lipid vesicle called a liposome. The self-reproducing system might be a collectively autocatalytic set of RNA, peptides, or both kinds of molecules.

Figure 6.1 sketches our hypothetical protocell. A collectively autocatalytic set is coupled with a metabolism of small molecules whose products include lipid molecules themselves. The lipids can enter the liposome shell, driving its growth. When it becomes large enough it buds into two liposomes—a primitive kind of cell division. Food enters from the outside through the semipermeable membrane forming the liposome. Similarly, waste is excreted.

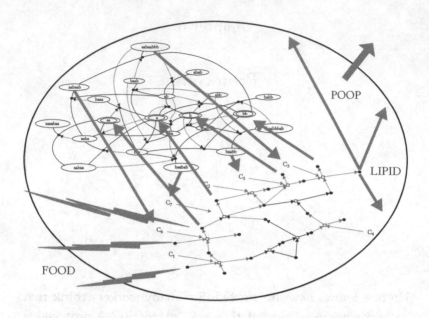

FIGURE 6.1 The protocell.

The Damer–Deamer Scenario

How might protocells arise? No one knows. There are two broad visions of where such an emergence might have occurred: hydrothermal vents in the oceans and fluctuating hot spring pools on land. Hydrothermal vents have been found to harbor rich simple life, and many hope that early life may have blossomed there. Computer scientist Bruce Damer and chemist David Deamer propose instead that the first protocells arose in interconnected pools of water on volcanic landscapes such as would have existed over four billion years ago resembling Iceland or Hawaii today (Damer and Deamer 2015). Evidence for such a hot spring rife with life in a 3.5-billion-year-old rock formation was recently discovered in Western Australia (Djokic et al. 2017). Wet–dry cycling—in which the water of the pool, or its edge, evaporates and recollects—occurs and, with an abundance of organic molecules, might have driven the process that I will sketch in the following.

A central part of the scenario pictures multilamellar, or multilayered, liposomes in a hot pool. Near the shores of this pool, wet–dry cycles would occur. As the day heats up, the pool dries by evaporation and refills in the cool night, as water seeps in from a nearby spring, or from rains. There are three stages: (1) When wet, the liposomes are hollow vesicles that float in the water; (2) when nearly dry, the liposomes form a gel-like aggregate among themselves; and (3) when dried onto mineral surfaces, they fuse together into layers, spilling their contents into the 2D spaces between the layers.

As wet–dry cycles follow one after another, this system cycles repeatedly between these states.

In their scenario, Damer and Deamer make use of the Plastine reaction, first studied in 1932. Large proteins are incubated with trypsin, a gut enzyme that during digestion cleaves large proteins into smaller ones. When a peptide bond is formed between two amino acids, one water molecule is released into the medium. Thus, if water is removed from the medium in such a system— for example, by evaporation—the reaction will thermodynamically *reverse*. New peptide bonds (and also nucleic acid bonds) can be formed, creating populations of polymers having initially random sequences. Now consider what happens in a wet–dry cycle: large polymers can be cleaved apart in an aqueous environment. Then when the system dries out, the parts of that polymer can ligate (join together) back into longer polymers. Repeated periods of wet and dry drive cycles of polymer formation from basic building blocks, cleavage, and re-synthesis—in which polymer fragments are randomly reshuffled with one another—are cleaved, and re-ligated, creating a soup of various polymers.

In the case of the Plastine reaction, if the trypsin catalyst is removed, the same thermodynamic forces are still at play, and the same reactions will happen, but more slowly. On the prebiotic Earth, there were no catalysts or enzymes, so it is reassuring that simple dehydration can do the same job, but more slowly. When water leaves through the stacked membranes layered in their "bathtub ring" on the mineral edge of the pool, the building blocks of polymers, already squeezed together by the sheets of membrane, line up and, like a zipper closing, form increased numbers of potentially functional polymers.

Thus, Damer and Deamer generate protocells, each housing a rich stew of different peptides or RNA sequences or both. During wet cycles, trillions of liposomes bud off when the layers of dried membranes bulk up with water. Some of these contain random sets of the aforementioned polymers, forming protocells. Cleavage will occur, during wet cycles, random shuffling and re-synthesis of peptide or RNA, but now this includes the peptide or RNA sequences inside the multilamellar liposomes. Together the liposomes and peptide soup inside them constitute the protocell that will ultimately evolve into the universal common ancestor

Damer and Deamer propose that as wet–dry cycles repeatedly occur over millions of years, a kind of natural selection comes into play—what Pross (2012) calls "Dynamic Kinetic Stability." Upon drying, each protocell crowds together with thousands of others that survived to this point, sharing their collective contents. Upon rehydration, the contents are captured in new protocells that bud off, and another cycle of cleavage occurs. Damer and Deamer suggest that those systems whose molecules somehow encourage stability will "survive" or "propagate" more efficiently, leading in due course to populations of increasingly robust protocells.

They refer to these vesicle populations as "progenotes," a term originally put forth by Carl Woese and George Fox (1977)

and adopted by Damer (2016). If they are right, such progenotes are the ancestors of all life on the Earth.

Part of what I admire about this scenario is that Damer and Deamer seem to have cleaved the Gordian knot of how to get something like heritable variation without having self-reproducing cells in the first place. If selection for Dynamic Kinetic Stability can happen in their system, they have achieved a form of variation and selection that might accumulate useful variant progenotes.

I wonder, however, just where this propagating stability that is selected might come from, and I now hope to build on their ideas.

Onward to Protocells

Damer and Deamer envisage selection for liposomes with *useful* mixtures of peptides and RNA-like polymers. But the random reshuffling due to cleavage and ligation will scramble these useful sequences in each cycle. Consider 10^3 useful peptides of length 10 amino acids in a liposome. There are 20^{10} or about 10^{13} such sequences. During each cleavage and re-ligation cycle, an initial 1,000 useful peptides will splay out into this sequence space at some rate, randomizing the useful peptides. It is not clear how heritable variation for useful polymers can happen, but it may arise first through template-driven replication.

I note next that hopefully the conditions envisioned by Damer and Deamer are just those that might give rise to collectively autocatalytic sets of peptides and nucleic acids within the gel aggregate or the multilamellar structure. So suppose such a set is stumbled on. And if the autocatalytic set is a stable reproducing set of polymers, then it could be selected for.

And that would give us a way to get the *same set of polymers over and over*. So suppose a collectively autocatalytic set forms inside a liposome. It dries out on the bottom or side of the pool, passing first through the gel phase, along with thousands of other liposomes, and then drying altogether. Somewhere along the way it fuses and spills its contents into the collective gel or dried layers, and the same polymers are taken up by its neighbors, which then rehydrate and begin the cycle anew. If any of these neighbors or the original one has a set of polymers able to regenerate the collectively autocatalytic set, that set proliferates to the neighbors in the gel or passes via the layered state to the newly forming neighbors rehydrating from the dried gel. In the next cycle, they can pass this property on to more neighbors. And so liposomes with "fit" autocatalytic chemistries proliferate.

Moreover, these systems should achieve dynamic kinetic stability. We have already seen that collective autocatalytic sets achieve the three closures: work task closure, constraint closure, and catalytic closure. These mean that *the same* system is recreated at each cycle. This proliferation constitutes dynamic kinetic stability. Compared to the random shuffling of polymers, progenotes with self-reproducing stable molecular systems in them will win and go on to cycle another day.

In short, the collectively autocatalytic set, by producing the same polymers, gives a selectable set on which selection can act and can also act on the help given to the multilamellar form to stabilize it in the environment—aqueous, gel, or dried— as it propagates. And the multilamellar forms help the set by confining and preserving their contents against the vicissitudes of the harsh Hadean environment.

A molecular mutualism is underway. This may allow selection not only for advantageous polymers in the set but for species of lipids in the propagating multilamellar progenote forms.

In chapter 5, we supposed that a collectively autocatalytic set could catalyze and graft to itself a connected catalyzed metabolism. The Damer–Deamer environment might just show how this could have happened on the early Earth—or anywhere in the universe with similar conditions. Such systems on the newly formed Earth would be rich in organic molecules, supplied by infalling organics as were shown to be present on the equally ancient Murchison meteorite, able to link into a catalyzed metabolism.

Some metabolisms would be better than others in supporting the reproduction of their associated collectively autocatalytic sets, so they would be selected and amplified through the Darwinian process.

Somewhere along the way, suppose that the metabolism came to form lipids as byproducts of no use to the autocatalytic set itself. Ultimately, such a union of a collectively autocatalytic set, a connected catalyzed metabolism producing lipids, and the capacity of those very lipids to form liposomes housing the autocatalytic set and metabolism could yield a form of "protocell mutualism."

One pathway to a more complex protocell might be the gel phase, where specific lipids could be of use to the local patch of protocells in the gel forming the multilamellar structures, so they might be selected. This might afford a co-selection of metabolism, lipids, and collectively autocatalytic set functioning well in the concentrating gel stage as the pool level drops.

Then one can envisage that these three "coupled phases" working together can now form a sophisticated protocell, with a boundary membrane, lipid synthesis, autocatalytic set, and metabolism as in Figure 6.1, able to divide by budding in the pool without dependence on wet–dry cycles and capable of free living in solution. At some stage in the "late progenote" world, protocells combine innovations until, accidentally (or fortuitously), they learn the trick of division, with all of their protogenetic and metabolic autocatalytic sets securely copied and divided between daughter cells representing—and voilà . . . . you have the transition to life (as we know it).

This brave scenario requires that the collectively autocatalytic set divide at the same rate as the housing liposome so they divide synchronously. Serra and Villani (2017) have shown that this arises easily.

Is this how life began? Maybe it is. It is all rather facile. Serra and Villani's recent book (2017) describes the difficulties of getting a protocell to work at all, given low concentrations.

At most, the preceding is a hopeful beginning and a call for further study.

Entropy and Persistent Self-Construction

A deep issue is how the biosphere builds up complexity in face of the second law of thermodynamics. This law states that in a closed system, disorder, or entropy, can only increase. In a system open to the input and efflux of matter and energy, entropy increases; but thermodynamic work can be done, building up complexity. In our hoped for protocell, lipids are built up. In plants, photosynthesis builds up glucose molecules from carbon dioxide and water. Fine, but if the second law degraded this order faster than it was created, no order could accumulate! How does order accumulate?

A sufficient answer to how order accumulates seems available to us thanks to the three closures: constraint, work cycle, and catalytic. In a system with constraint and work-cycle closure, the constraints on the release of energy in the non-equilibrium processes does work, and this work is used to construct yet more of the same constraints. This is the harnessing of energy to build up further order. That these systems are

also parts of reproducing molecular systems, due to autocatalysis, means that they can reproduce and further construct
order faster than the second law dissipates that order. There is
a persistent self-construction. As we see soon, such systems
can evolve by heritable variation and selection. A biosphere
can now build itself.

Things Not Said

Getting to protocells from dumb, dead earth is already a lot.
But current life is based on something more: DNA that codes
for the production of proteins including the ones, like DNA
polymerase, that the DNA needs to replicate itself and very
much more (prokaryotes, eukaryotes, multicellular organisms,
sex . . .).

   We can imagine the first steps, but the mystery is huge.

Chapter 7

Heritable Variation

Darwin was right. With heritable variation and natural selection, along with some form of organized propagation such as we seek in this book, the glory of the diverse biosphere can and did arise. Bumble bees, redwood trees, sea urchins, rooks on rocks at low tide: we live in a flowering forth.

In contemporary cells, heritable variation arises by mutation and recombination of genes and is based on encoded protein polymerase replication of the helix of DNA in cells. But this takes genes, encoded protein synthesis, and more. Such were not present in early life, whose adaptive evolution by heritable variation and natural selection was needed to arrive at genes and encoded protein synthesis in the first place.

Then how can protolife, perhaps collectively autocatalytic sets or nude replicating RNA in liposomes, show heritable variation?

If protocells harbored replicating RNA ribozyme polymerases, these would be capable of heritable variation by miscopying themselves. The problem is that they would be subject to the Eigen–Shuster error catastrophe that we saw earlier in the book. That is, at a low mutation rate, each population of RNA sequences would hover in the vicinity of its master sequence in sequence space. As the mutation rate rises, the population would rapidly diverge, and all sequence information would be lost. That would be even worse if the erroneous copies of the polymerase were more error prone than the wild type so that the mutation rate increases with each replication cycle, as discussed before. The useful sequences would melt away.

If the polymerase were able to gather around itself other RNA sequences of functional importance to the protocell, these would be reproduced at a constant mutation rate, and so also be subject to the Eigen–Shuster error catastrophe.

We cannot know the mutation rate of an unknown and still to be found replicating RNA sequence, so the issue is a bit moot. If the rate is low enough, the system can evolve; if too high, it cannot.

If the protocell houses a collectively autocatalytic set, it can act as the unit of inheritance. Typically, this would consist of one or many irreducible autocatalytic loops with a tail that plays no role in autocatalysis. As Vassas et al. (2012) point out, the loop is the analogue of a gene; the tail is an early phenotype coupled to that gene. That is, the tail is not needed for reproduction of the set, so can play other functional roles, such as catalyzing the metabolism. Collectively, autocatalytic sets can evolve by gaining and losing irreducible sets that act as genes.

In the Damer–Deamer scenario, it may be relatively easy for collectively autocatalytic sets to emerge. A liposome is about a micron across. So in a 10-micron-square area some 100 protocells could be found forming into or emerging from a gel. In a square meter, that becomes 10^{12} protocells. As Damer and Deamer note (2015), there are very many possibilities for local experimentation, including exchange of irreducible autocatalytic set loop genes.

As we saw in chapter 6, it is easy to imagine protocells in the Damer–Deamer wet–dry scenario gaining and losing irreducible autocatalytic sets. Two liposomes readily fuse, allowing the sets to fuse into a new union, thus sharing their irreducible sets. And liposomes can bud, allowing irreducible sets to be lost from one of the daughter liposomes by random diffusion to these daughter cells.

Additional means exist for collectively autocatalytic sets to evolve. As Bagley and Farmer noted years ago, the components of the set are present in relatively high concentration and therefore can drive spontaneous reactions for new molecular species for which they are substrates. If those products become glued reciprocally and catalytically to the set, it evolves new molecular species.

Finally, catalysis is not totally specific, so any polymer may catalyze a cloud of similar reactions, providing variability.

And so we have taken another important step: collectively autocatalytic sets can evolve by heritable variation and natural selection. Housed in liposomes, they become units of selection. Our progenotes, the ancestors of us all, can evolve.

Chapter 8

The Games We Play

I began chapter 2 by posing a question: How since the Big Bang did the universe get from matter to mattering? The answer in brief is once there are protocells evolving, mattering matters.

In the late 1990s, I found myself struggling with the issue of agency. What must a physical system be to be an agent—a doer—able to act on its own behalf? I found myself considering an answer (Kauffman 2000): a molecular autonomous agent is a system that can reproduce itself and do at least one thermodynamic work cycle. For example, consider a bacterium swimming up a glucose gradient. The sugar matters to the bacterium. Mattering is now part of the universe. Agency introduces meaning into the world! Agency is fundamental to life.

I know of no way to derive my definition. Definitions are odd things in science—neither true nor false, but hopefully useful. Newton's F = MA is, as Poincaré pointed out, a definitional circle. Force is defined in terms of mass and mass in terms of force via an independently defined quantity called "acceleration." Yet the definition carves nature at the joints as seen through the power of classical physics such as celestial mechanics.

Biologists worry that Darwinism is circularly defined, as survival of the fittest with the fittest defined as those that survive. Yet Darwin unpacks the biological world for us. It may be that my definition of agency is fruitful, for, as noted, definitions are neither true nor false. Nevertheless, definitions allow us to see the world in new ways and can be deeply useful.

An agent doesn't have to "know" it's an agent. Based on our definition, Ashkenasy's nine-peptide collectively autocatalytic set (Wagner and Ashkenasy 2009) is already an agent, for it reproduces and does a cycle of work. The only difference between a cycle of work and a thermodynamic work cycle is that the latter must also couple a nonspontaneous endergonic and a spontaneous exergonic process. This is of little matter, as it is easily fixed. In *Investigations*, I showed a simple self-reproducing system that does a real thermodynamic work cycle.

Ashkenasy's set is not yet housed in a liposome. It is not a protocell. Once that step is taken, all doubt is removed. I am perfectly happy thinking of such a system as a rudimentary molecular autonomous agent.

Sensing, Evaluating, Responding to the World

Consider evolving protocells, in and beyond the Damer–Deamer pool of chapter 6, able to reproduce and evolve by whatever means. Consider the selective advantage of being able to sense one's world, the presence of food, the presence of poison—to be able to evaluate it, "yum or yuck, good or bad for me"—and to be able to respond in some way to that environmental situation, choosing food, avoiding poison.

The selective advantages for the emergence of this ability would have been enormous. Consider it achieved. Mattering has evolved: this is "good or bad for me."

Katherine P. Kauffman[1] (personal communication, September 2017) considers the preceding triad—sensing the world, evaluating it, and acting on it—to be the foundation of emotion. I suspect she is right and that, as she argues, emotion may be the first integrated "sense." From it and integrated with this first sense, all other senses would have evolved, coupled with evaluation—good for me, bad for me.

Moving

So far our protocell cannot crawl, let alone walk. But we can fantasize the evolution of a capacity to do so. This might arise by control of an internal "sol-gel" transformation, from a wet slurry to a jello-like state—for example, by chemo-osmotic pumps—such that one part of the protocell controllably becomes a liquid like sol and the other a firm gel, allowing the sol region to move with respect to the gel region, and hence leading to amoeba-like movement.

For Plato and the Ancients, self-moving is one sign of "soul." We have the rudiments of soul and vitalism. This is the transition from the inanimate to the animate world.

From Matter to Mattering

Agency has arrived. With agency comes doing. The protocell, or bacterium, acts to get food and avoid poison. However the acting

Peil, Katherine T. "The Emotion: The Self-Regulatory Sense." 2014. *Global Advances in Health and Medicine*. 3(2): 80-1-8.

is done, for example, swimming with a flagellum for a modern bacterium, or sol-gel amoeboid motion for the protocell, this is not a mere happening but a doing.

Why do we make this distinction? In chapter 2, we discussed functions as real in the universe by virtue of the role that some causal aspects of organisms play (the heart pumping blood) such that the organism gets to exist in the nonergodic universe above the level of atoms. When the agent goes to get food, it performs a function. And so it too gets to exist in the nonergodic universe above the level of atoms. The getting is a function—a doing and not a mere happening.

Instrumental Ought

David Hume said one cannot deduce an ought from an is, the famous naturalistic fallacy. From the fact that mothers love their children, you cannot deduce that they ought to do so. But Hume is deeply wrong. Hume, in the tradition of British Empiricism, thought of a passive observing mind/brain in a vat and wondered how that observing mind could have reliable knowledge of the world. He rightly noted that from what is observed to be the case, one cannot deduce what ought to be the case. We live with the naturalistic fallacy.

But Hume forgot that organisms act, even one as simple as an amoeba. Once doing enters the universe, doing it well or doing it poorly follows. Trying to eat my ice cream cone, I might keep bashing it into my forehead. In short, doing brings with it instrumental ought. We are agents, and doing something well or poorly matters! So we ought do it well. This is instrumental ought, not moral ought. Instrumental ought, how to do something, is with us as soon as agency arrives. Thus, it is ancient.

The Filigreed Games We Play

Rocks cannot evolve to avoid or deceive other rocks. But agents can. Once there is food and poison and simple food chains, the prey can mimic being poison to hide from the predator. Mimicry is all over the place in evolution. Some butterflies mimic the markings of other butterflies that taste bad to avoid predation. Once there are food webs, the prey can evolve defensive procedures in face of the predator. With and without food webs, in ecosystems, the games organisms evolve to play with one another are myriad. Each time, meaning sprouts again in new myriad ways.

The games we play with one another "are us together getting to co-exist and get on with us getting on with life in the nonergodic world above the level of atoms, abet the unending invention of further games." We play filigreed games. Moreover, we are such that we can have evolved to play such games. We create interwoven webs of collaboration and competition in untold ways. As we will see, this evolution is not prestatable. Not only do we not know what will happen, we do not even know what can happen! This topic will be with us in the rest of this book, for it underscores the unprestatable evolution of the biosphere—and even the evolution of our global economy.

The Surprising True Story of Patrick S. "The First," Rupert R., Sly S., and Gus G., Protocells in Their Very Early Years

The Story of Patrick!

Once upon a time, very, very, very long ago, almost four billion years ago, off the West Coast of Gondwanaland, life as protocells had recently begun. It all happened under a turbid sun, on a scorched Earth, in a shallow lagoon. Days and nights came and went, before Patrick, Rupert, Sly, and Gus, really became Patrick, Rupert, Sly, and Gus. For now they were merely protocells, undistinguished and ordinary amid their generation X cousins, dry and wet and wet and dry, and all the Gen Xers passively absorbed the stuff softly flowing in the lagoon. You could sort of call it eating. And they multiplied, making so many Gen Xers that almost four billion years later their grand, grand, grand, grand, . . . you know, us and others, would be all over the blue dot planet.

But back then nobody got much "stuff" to eat because all of the even tinier floating stuff floated at almost the very same speed as did the Gen Xers. That was okay because it was true for all of them, and nobody got really mad.

But one day, Patrick Protocell felt a bump jump hurt inside himself. "What's that?" he thought, a bit fearfully. "Oh it's my whatdyacallit sticking out! Ouch."

Patrick felt the pinch and was even pierced. A little molecule, a peptide made of thirteen amino acids, protruded from his side.

Then do you know what happened? This little peptide bumped into a huge rock, very much bigger than Patrick, but much smaller than even a thimble.

And the peptide stuck to the big rock. Patrick himself was *stuck*. He could not float about and laugh in the lagoon, hoping for stuff.

"I've got to get unstuck," thought Patrick with alarm. He yanked his tummy and his bottom up, but stayed stuck. The more he tugged the more stuck he seemed to be.

"Oh NO!" thought Patrick. "All is lost. If only I had a mother I could call her!" he winced.

"Oh well, maybe I'll get unstuck when it gets wet and dry a few times," he hoped, rather like a latter-day sailboat hung up on a rock at low tide.

"I'll have to make the best of it 'til then."

"Maybe I'll still bump into some stuff," he hoped.

"But *how*, I'm all stuck on this old rock!"

Patrick, without too much hope, a bit desperate at his woeful situation, looked up, and guess what?

Well, you'll never guess what happened to Patrick.

In a trice, Patrick saw flooding at him lots of stuff, more than he had ever seen, here and there, everywhere, floating so fast toward him he feared he could never ever gobble it up.

So, bucked up by the very possibility, Patrick gobbled as fast as he could.

Very full, a very short time later, much shorter than usual, Patrick divided into two Patricks.

"We're stuck," they both cried. And indeed they were—to the very same big huge rock.

Patrick and Patricks were dividing so fast now and they had so much stuff flowing at them that soon there were lots of Patricks!

In about seven moons there was a large Patrick Patch, many grandchildren of Patrick, who had become, what?

Patrick had become, on getting stuck to the huge rock, the very first "sessile filter feeder" on the early planet Earth. Think of that. The very first one.

And that is how Patrick became PATRICK THE FIRST!

Before Patrick got stuck, he was a typical sophomoric Gen Xer protocell. Now he was special. He could stay stuck to the rock, sessile filter feeding all wet and dry long.

Where had Patrick come from? Well, sort of from nowhere! Patrick the First just emerged!

First there were just Gen Xers, Patrick among them. All slowly dividing while sort of eating stuff.

But Patrick had, accidentally of course, seized a special opportunity. There were nutrients flowing slowly and there were rocks, including the rock he got stuck to. So *if* he did get stuck, he'd get more nutrients per unit time than other protocells, and so divide faster.

But what does it take in the becoming of the universe for a context, like the rock and slow-flowing nutrient stream was to Patrick, to be an "opportunity?" For the rock, the flowing water is a context but not an opportunity.

Not everything or every process is an opportunity. A tiny rock by itself is not an opportunity. Nor is a rock and a slow-flowing stream of stuff. There is no opportunity without something that can seize that opportunity and take advantage of it.

And Patrick was just such a "something." Patrick had seized his opportunity, "For ME" thought Patrick, glad that he was one for whom an opportunity of a lifetime could be seized.

Patrick had become a "for whom."

What does it take for something in the universe to "seize an opportunity?"

What does it take for something to become or be an opportunity that can be seized and for something to be able to seize it?

The surprising crux of the matter is worth repeating: you cannot have an opportunity without something, a "for whom," for whom the context *is* an opportunity that now can be seized.

What counts as an opportunity makes no sense without something that can seize the opportunity. But this is not imaginary and not just words. Patrick really came to exist in the early biosphere as the first sessile filter feeder; hence, he came to exist in the nonergodic universe above the level of atoms by seizing his opportunity. He became Patrick The First Sessile Filter Feeder.

What counts as having seized an opportunity? For Patrick and the biosphere, the success was very real: More Patricks forming the Patrick Patch did in fact outgrow the Gen Xers.

Patrick and his offspring could do this because they were autopoietic, that is, they were self-reproducing systems able to self-maintain and reproduce, have heritable variation, and be selected. Thereby, he and his offspring could seize his opportunity. He and his offspring were Kantian wholes where the whole exists for and by means of the parts.

In particular, Patrick was a collectively autocatalytic set of peptides in a liposome, a hollow lipid vesicle that buds and is also able to make the lipids to form the liposome. Patrick was an early form of life able to evolve by *heritable variation* and natural selection. That is why Patrick constituted a "for whom" so that a context, here the slowly flowing nutrient stream and the tiny rock, constituted an opportunity to be seized. Patrick came to exist in the nonergodic universe above the level of atoms, where most complex things will never exist. Patrick actually changed the unfolding history of the whole universe. No mean feat when all he had to hold onto was a tiny rock not as big as a thimble.

"I'm so glad," thought Patrick The First. "I'll just hang in here and love it and divide when I feel like it."

So Patrick divided and made lots of Patricks, two by two, until before he knew it the Patrick Patch had spread over a big part of the lagoon.

That is the first part of Patrick's story, how the first sessile feeder came to exist out of pretty much nothing.

And the story is all you need to know. That's really what happened. Isn't that just amazing? First no Patrick, then Patrick "The First," sessile filter feeding, out of nowhere. Just because his peptide happened to stick to the rock.

Later, Darwin would call this sticking a preadaptation in Patrick.

The Story of Rupert!

And now Rupert's story (how Patrick, now that he exists, provides an opportunity for Rupert to emerge and exist).

Rupert was pretty much your ordinary protocell, although a bit laconic, somewhat more laconic than the others. He could not swim but could wiggle a bit as he came near stuff. Maybe he was excited, so he wiggled. But beyond wiggling, Rupert was already a bit special. He could eat stuff, but he could also stick to other Gen Xers and make a hole in them and suck out their inside stuff. Rupert thought this was good, for every now and then he bumped into another Gen Xer and got a special dinner from it.

But bumping into other Gen Xers did not happen very often, as they were all floating in the same slowly moving stream of stuff. Rupert, like the others, mostly ate plain old stuff.

One day, do you know what happened? Rupert floated into the Patrick Patch far away from most of the lagoon.

"OH NO," thought Rupert, "This place is full of. . . . Well I don't know. How do I get back to the clear lagoon?"

He tried wiggling but got nowhere fast. It was the best he could do.

Rupert was as woeful as Patrick had been, maybe more. He was far from the clear lagoon.

Guess what happened to Rupert? He bumped into Patrick the MMMMCCCDXXXVIII!

Rupert poked a hole in that sad Patrick and ate him up.

"GHA," thought Patrick the MMMMCCCDXXXVIII.

"Cool," thought Rupert.

So Rupert became famous in the lagoon as Rupert "Raptor" Protocell. He was the very first predator in the lagoon and on the whole earth and in the universe. Rupert changed the history of the whole universe.

Soon there were lots of Ruperts bobbling in the Patrick Patch, which itself was growing in the number of Patricks faster than the Ruperts could manage to eat them all. This was the very first food chain in the biosphere. Out of nothing did it come. The first food chain changed the history of the universe. (So will the rest of the food chains that followed.)

Rupert, like Patrick, was a "for whom" so there could be an opportunity. The startling thing about Rupert, however, is that Rupert's opportunity included not only the lagoon with nutrients but now included Patricks. Because Patricks were sessile feeders stuck to rocks, Ruperts bumped into Patrick and his kin far faster than into Gen Xers floating in the nutrient stream in which Rupert and his kin also floated.

Patrick was *part of the whole context* that was Rupert's opportunity. Rupert seized his opportunity. Patrick, by existing
and creating a Patrick Patch, *afforded* an opportunity to Rupert,
given that Rupert Protocell could not swim and was in a slowly
moving nutrient stream where he could only eat stuff—and,
very rarely, bump into Gen Xers. So Rupert's opportunity was
Patrick the First and kin, the sessile filter feeders in the Patrick
Patch where Rupert could bump into many of Patrick's kind,
compared to just eating stuff and the occasional rare treat of
eating a Gen Xer.

With all this stuff to eat, Rupert now divided rapidly; and
soon there were lots of Ruperts growing in the Patrick Patch, or
what by now were several Patrick Patches in the lagoon.

There was no one else alive in Patrick's context of opportunities. Patrick's opportunity was only the slowly flowing stuff
and the tiny rock he sort of grabbed onto. But by coming to exist
in the universe, Patrick and his own kin in the Patrick Patch now
came to constitute the "context," the very opportunity for Rupert
to come to exist: no Patricks, no Ruperts, who soon quite forgot
about eating the hard-to-bump-into Gen Xers and now depended
entirely on eating Patricks to survive.

The ecosystem had become Gen Xers, floating stuff, Patricks
in Patrick Patches, and Ruperts grazing on Patricks. This is a bit
like, billions of years later, grass and rabbits.

Could you write an equation for this? How would you know
what to write? This story is pretty much what you need to know.
What would mathematics do here at all? It could not do much
about the becoming of Patrick and Rupert. In fact, mathematics
would tell us nothing about this becoming.

But Pythagoras taught that all was number. Is it? Where is
the "number" here? We, looking on, do not need number. And
Patrick and Rupert never heard of Pythagoras who grazed in the
Agora long thereafter.

The Amazing Story of Sly Protocell

To start with, Sly was pretty much an ordinary protocell except he could, like the early Rupert, eat Gen Xers if he happened to bump into them, as well as eating the floating stuff.

Sly, who did not know that his name was rather pejorative, was perfectly happy. He floated in the lagoon and ate.

One day, Sly bumped into a Rupert. And do you know what happened? By accident, a peptide on Sly's surface attached to Rupert! Sly was embarrassed, and Rupert was annoyed at this bondage. But the choice seemed to be Sly's. Rupert could not shake Sly off.

And what do you think happened?

Now when Rupert ate a Patrick, some of the juice squeezed out of Rupert's insides through the hole, and Sly licked up the leftover juice from Patrick's perishing.

Actually, Rupert came to be glad about the arrangement because the juice on his outside self felt sticky. Sly was a bit like small fish inside a shark's mouth cleaning the shark teeth. It's a strange way to make a living, huh? But Sly changed the universe because Sly divided faster than before, and soon there were lots of Slys attached to lots of Ruperts all over the Patrick Patches in the lagoon.

But Sly did more. You see, Patrick and his offspring had trouble attaching to the tiniest rocks, much, much smaller than a thimble, and sometimes slipped off. But when Sly slurped up the juice from Rupert's gobbling a Patrick, Sly seemed to excrete a glue into the little area of the lagoon that helped glue Patrick to the rocks! So in the presence of Ruperts with Slys, Patricks lived more securely in their Patrick Patches, more firmly attached to the tiny rocks.

What had come about? Sly had come to exist. His opportunity consisted now of both Ruperts and Patricks. Sly was also a "for whom" who seized his opportunity. Now Sly existed too, out of nothing much.

But there is more. Rupert no longer ate Gen Xers, as told before. But Patricks slipped off their rocks sometimes and died, lowering the number of Patricks on whom Ruperts could graze, therefore holding down the population of Patricks and also of Ruperts. But Sly helped glue Patricks to their rocks more firmly, so everyone benefitted. Patrick provided a niche for Rupert, who provided a niche for Sly, who helped provide a niche for Patrick! They formed a three-species, "collectively autocatalytic set"! Such collectively autocatalytic sets of species mutually creating niches for one another exist today.

In fact, the Sly glue was so great that Patrick sort of forgot how to attach very well to rocks and now depended pretty much entirely on Sly. The autocatalytic little ecosystem became tighter and mutually co-dependent. They worked well together, and Patrick and Rupert and Sly and their kin got to exist in the nonergodic universe for a pretty long time.

The Story of Gus

Gus was also just your ordinary Gen Xer. He bobbled around in the lagoon like all the rest.

Every now and then he saw a tiny rock and reached for it, but he could not grab the rock; so he floated and divided, but not very quickly.

One spring day, Gus bobbled into a Patrick Patch, and guess what?

Gus bumped into a Patrick and Gus found that he *could* grab a Patrick. And so he did.

Guess what he learned?

Gus was indirectly stuck to Patrick's rock! He was quite glad, for he had tried and failed before to grab a rock on his own. But now the slowly flowing stream of stuff that floated rapidly by Stuck to Gus too, and he ate lots more stuff. Like Patrick, Gus divided faster. Sometimes there were two or three Guses attached to one Patrick, who was rather annoyed but could not shake Gus off because Patrick could only wiggle.

Gus is a "for whom" and Patrick is his opportunity. So Patrick afforded *two* new niches, constituted two new opportunities, one for Rupert, and one for Gus!

Darwin once described an image of a species driving a wedge in the crowded floor of nature—a wedge of a competitive nature that created a space for it to live in. That is not the story of Patrick, Rupert, Sly, and Gus. Patrick, in seizing his opportunity and becoming Patrick "The First" and forming the Patrick Patch, thereby created and afforded a new niche for Rupert. Patrick *is* the niche and opportunity for Rupert. Rupert *is* the niche for Sly, and Sly with his glue becomes part of the niche for Patrick. And Patrick is the niche for Gus.

There is no wedge driven into the crowded floor of nature. The floor itself is expanding, creating new niches by creating Patrick, Rupert, Sly, and Gus—who create the niches for one another. Patrick, Rupert, Gus, and Sly create new cracks in the floor of nature, new niches, for one another. The same is largely true of the biosphere, and the global economy, both of which have exploded in diversity just as Patrick gave rise to Rupert who gave rise to Sly who stabilized the three-species ecosystem to which Gus came along to hang off Patrick.

We seem to make our worlds and thereby make rooms for one another. Each "for whom" makes even more opportunities for others in its adjacent possible niches or rooms. The adjacent possible niches, like worms coming to live in swim bladders, explode faster than the number of occupants who, by existing, create those very adjacent possible niches.

In much the same ways, both the biosphere and global economy explode in diversity. Each species affords one or more adjacent possible new niches for yet new species, which so expands what now becomes possible. Spanish moss hangs from laboring trees. New goods and services and production capacities expand the ways in which further new goods and services can now make a living. Personal computers made word processing possible, which made file-sharing possible, which made the World Wide Web possible, which affords a place to sell on the Web, which made content on the Web that soon enabled browsers. The introduction of the automobile enabled the gas industry and paved roads. Paved roads required traffic control. The roads enabled motels and fast food restaurants.

It is not only that the floor of nature is crowded by competition, as Darwin thought, but rather that each species also affords adjacent possible new niches, new "wide cracks" in the floor—new niches that invite the next new species into those wide cracks that constitute new niches. The possible new niches expand faster than the species that create them. Patrick created two niches, one for Rupert, one for Gus. The Web enabled both eBay and Amazon.

This is an unprestatable becoming of "for whoms" that can seize their specific opportunities in adjacent possible niches that we each in turn create. The "floor of Nature" expands, housing ever more room after room that we jointly co-create faster than we all come into existence. And that is how complexity emerges.

Chapter 9

The Stage Is Set

With Patrick the First and his friends, the stage is set. Life has started, and the biosphere will flower forth. Patrick, Rupert, Sly, and Gus are protocells, unleashing an unprestatable becoming. They emerge and evolve in a lagoon similar to the one Damer and Deamer (2015) envisioned. They adapt by what is called Darwinian preadaptation—that is, they have properties that were not selected to perform a given function but that can take on that function if the opportunity arises. Feathers evolved for thermoregulation and were co-opted for flight. Patrick, for example, had a peptide that stuck out of his insides, having evolved for some other function or for no function at all, but it happened to be able to stick to a rock. And so Patrick became Patrick The First Sessile Filter Feeder.

We will talk about Darwinian preadaptations more in chapter 10, for although we cannot prestate them, they drive much of evolution. Patrick, by sticking to the rock, gets more food per unit time, and so a new kind of "species" is born. The rock is Patrick's opportunity, and Patrick is a "for whom" and so can benefit from an opportunity "seized" by heritable variation and natural selection. There cannot be an opportunity without a "for whom"—*for whom* it is an opportunity.

Such evolving early forms have the astonishing property that each, by coming into existence, can constitute a new "context" and opportunity that does not cause but "enables" yet further life forms, "species," to come into existence. Patrick constitutes the empty niche into which Rupert becomes. Rupert and Patrick jointly constitute a new niche enabling yet other life forms, Sly and Gus, to come into existence in ways that "depend on" Patrick and Rupert as prior existing conditions. The increasing diversity of organisms and niches in turn affords yet more opportunities for further "species" to emerge. And this creates yet more contexts that afford yet more opportunities.

As in the story of Patrick and friends, Darwin had an image of species driving wedges into the crowded floor of nature to make room for themselves. But there is more. Patrick, by existing, *constitutes* the new crack in the floor that allows Rupert to come to exist. Rupert does not have to make a crack for himself. Patrick *is* the crack. Rupert is the crack for Sly, and Patrick is the crack for Gus. As the diversity of species goes up, the number of new cracks, new adjacent possible empty niches, goes up faster than the number of species. Diversity explodes! Patrick supplies two niches, one for Rupert, one for Gus. The World Wide Web supplied myriad new niches, including one for eBay and one for Amazon. The global economy explodes in diversity. By coming to exist, species create the adjacent possible opportunities for other species to make a living in new ways.

The Biosphere Explodes in Diversity

Richard Dawkins has famously written *The Selfish Gene*, stating that evolution is a more or less brutal race for the survival of genes, and further, that organisms are merely the vehicles that carry the genes to be selected. But the story is deeply inadequate. As we just saw, Patrick, by existing, constitutes the empty niche into which Rupert can become; Rupert constitutes the niche into which Sly can become; and Patrick constitutes the niches into which Gus can become. Species, by coming into existence, literally create new niches into which other species come to exist. Moreover, the niche does not cause the new species to come to exist but rather affords an opportunity that *enables* the new species to seize that new niche and further evolve.

How vastly richer than "nature, red in tooth and claw." It is not Patrick's genes that are selected, for Patrick has no genes at all in the familiar sense of DNA. There are no selfish genes; there is the whole organism called Patrick. Dawkins has forgotten the organism. Organisms are selected, genes go along for the ride. We will see the same niche creation in chapter 10 with other Darwinian preadaptations. New species create new niches for yet further new species. Now there are millions of them.

We Cannot Mathematize This Becoming

I told of Patrick, Rupert, Sly, and Gus as a children's story. Could I have written equations and derived the becoming of Patrick, Rupert, Sly, and Gus from your ordinary "Generation X" protocells? Well, no, I couldn't. Try to do so. Just what variables would you write down? How would you simulate this emergence on a computer? I see no way, do you?

Pythagoras taught us that all is number. Newton, after Galileo, taught the same thing. Nature is written in number, the "rule and line" that Keats decried. It is a major transformation in our way of knowing the world if we can write no equations for the emergence of Patrick, Rupert, Sly, and Gus. We understand the children's story perfectly. We tell it as a narrative. What else can we do?

This will be a major theme of the rest of this book. We cannot derive this becoming by equations. The becoming is not derivable by entailing law, for we can write no laws of motion for the evolving biosphere, as we do not know the relevant variables prior to their emergence in evolution. We do not know that Patrick will stick to the rock by the peptide that protrudes from his insides. We cannot mathematize the specific evolution of the biosphere. We can, at best, seek statistical laws about distributions of aspects of this evolution. In short, I will claim that no law at all entails the becoming of the biosphere; and that therefore, we cannot reduce biology to physics. The world is not a machine.

Context-Dependent Information

Patrick, Rupert, Sly, and Gus develop context-dependent information about one another. Rupert comes to know Patrick and Patrick's habits. For example, Patrick learns to "crenelate," or crunch up, to avoid being eaten, but Patrick can only hold that pose for a while, and Rupert can still eat Patrick some of the time if Rupert waits cunningly. In short, Patrick, Rupert, Sly, and Gus start to play context-dependent "games" with one another. Organisms can evolve to play games of increasing diversity: filigreed games. Rocks cannot do this. Touch a clam snout and watch it spout in the sand. With the diversification of the blossoming biosphere, context-dependent information explodes.

And so the stage is set. From the lagoon, life springs forth. And thanks to the three closures—constraint task, work task, and catalytic task—life physically constructs itself and literally surges upward in complexity in the nonergodic universe above the level of atoms. This surging is A World Beyond Physics, the title of this book.

Chapter 10

Exaptations and Screwdrivers

Can we say ahead of time what will arise in evolution? The burden of this chapter is that, often, we cannot. We cannot prestate what will arise, just as we could not have prestated Patrick's emergence as the first sessile filter feeder.

Preadaptations and Exaptations

We have noted more than once that the function of the human heart is to pump blood but that the heart also makes heart sounds and jiggles water in the pericardial sac. Were Darwin asked why the function of the heart is to pump blood, he would respond that it was selectively advantageous in our ancestors to have a heart that pumps blood and that by virtue of this causal consequence, the heart has been selected and passed down to us.

Darwin had many brilliant ideas. Among them was his further realization that the heart, in a different environment, might come to be selected for some causal aspect other than pumping blood. Perhaps as a resonant chamber, my heart might pick up earthquake pre-tremors. I rush outside, escape a dreadful quake, become famous, mate with many, and my dominant genetic mutation for a heart able to sense pre-tremors is passed on to my manifold children. Well, it's not likely, but conceivable.

In short, Darwin realized that a causal consequence of some aspect of me, of no selective significance in the current environment, might become of use in a different environment and so be selected. With that, a new *function* would come to exist in the biosphere. These are very common and are called Darwinian preadaptations, with no hint of foresight on the part of evolution. S. J. Gould renamed these Darwinian exaptations.

Exaptations are indeed common. Your middle ear bones, incus, malleus, and stapes, evolved as exaptations from the jawbones of an early fish. Presumably these were sensitive to sound vibrations and co-opted for a different use. Feathers evolved for thermoregulation but were co-opted for a different function, flight. The well-known flagellar motor, used by many bacteria to swim, was not assembled at once. Rather, its protein components were of other uses and co-opted for use in locomotion.

Perhaps my favorite Darwinian preadaptation is the swim bladder. Some fish have a bladder that holds air and water. The ratio of air to water in the sac tunes neutral buoyancy in the water column. Paleontologists think swim bladders evolved from the lungs of lung fish. Water got into some lungs, which now had a mixture of air and water and were poised to evolve into a swim bladder.

With the swim bladder's emergence, a new function came to exist in the biosphere: neutral buoyancy in the water column.

But there is more. Like Patrick affording a new adjacent, possible empty, niche for Rupert, might a worm or bacterium evolve to live only in swim bladders? Yes, of course. So the swim bladder, by existing, opens a new crack in the floor of nature, to borrow from Darwin, and a worm can live in that new crack.

And there is still more: Does the bladder *cause* the worm to evolve to live in the swim bladder? No. The bladder enables the worm to evolve to live in the swim bladder—a subtle but crucial difference.

"Enablement", not "cause," enters our explanatory vocabulary! In 2012, Longo, Montévil, and I published a paper, "No Entailing Laws, but Enablement in the Evolution of the Biosphere" (Longo, Montévil, and Kauffman 2012). All of the niche creation of which we speak in this open-ended evolution is enablement, not cause. This can be seen in more detail. The mutation in the worm that is part of the evolution of the capacity to live in swim bladders is itself a random quantum event. Much of the becoming of the biosphere has to do with *making possible*. The same is true of the evolving economy as discussed in the Epilogue.

Natural selection played a role in "fashioning" a working swim bladder. But did natural selection fashion the swim bladder such that it constituted an adjacent possible empty niche in which a worm could evolve to live? NO! But that means that without selection accomplishing it, evolution creates its own possibilities of future evolution! Evolution, without selection achieving it, evolves its own future pathways of becoming!

Do you think you could have said ahead of time that the swim bladder—like Patrick as the world's first sessile filter feeder—would emerge? Could you have *prestated* the swim bladder, flight feathers, bones in the middle ear, Rupert and Sly and Gus? No. Try to prestate all the Darwinian preadaptations in humans for the next 5 million years. You cannot. We'll see why in a moment when we discuss screwdrivers.

But this means something vast: not only do we not know what *will* happen, we do not even know what *can* happen. Compare this to flipping a fair coin 1,000 times. Will it come up 540 heads? We do not know, but we can calculate the probability via the binomial theorem. We do not know what will happen. But we know what can happen—all 2 to the 1,000th power possible outcomes of 1,000 flips. We know the sample space of the process. For the evolution of the biosphere via exaptations, we do not know even that! We do not even know what *can* happen.

But this means we can formulate no probability measure on what happens, for we do not know the sample space.

It will follow from this, later, that we can write no laws at all for the specific evolution of the biosphere, whose becoming is thus entailed by no law, so the evolving biosphere is not a machine.

The Many Uses of Screwdrivers

I hand you a common screwdriver. Please list for me all the uses of a screwdriver in, say, New York, in 2017. Well, go ahead: screw in a screw, open a can of paint, scrape putty from a window, stab someone, display as an objet d'art, scratch your back, wedge the door open, prop a window open, jam a door closed, tie to a stick and spear a fish, rent the spear out at 5 percent of the local catch, and so on.

Is the number of uses of a screwdriver infinite? No, for discretely different things, like uses of screwdrivers, to mean "infinite," we require a recursion, 0, 1, 2, 3, N, N + 1 as for the integers. . . . But if we have N uses of a screwdriver, what is the next, N + 1, use of a screwdriver? Can you enumerate it forever, for all N to infinity? No, you cannot.

The number of uses of a screwdriver is *indefinite*. Will you accept "indefinite"? If so, your life is lost.

I now remind you of four levels of scales: (1) a nominal scale is just a set of names of things, with no ordering relation among the members of the set; (2) a partial ordering scale is one where X is greater than Y and Y is greater than Z, so X is greater than Z; (3) an interval scale is like a thermometer, where the distance from 0 degrees to 1 degree is the same as from one to two degrees, but zero means nothing; and (4) a ratio scale is a meter stick, where two meters is twice one meter.

The uses of a screwdriver are merely a nominal scale. There is no ordering relation between the different uses of a screwdriver and no fixed intervals.

I claim two major results: (1) *no rule-following procedure, or algorithm, can list all the uses of a screwdriver*; and (2) *no algorithm can list the next new use of the screwdriver!*

I believe these claims are correct. *We* cannot state all the uses of a screwdriver nor deduce the next new use of a screwdriver.

But Darwinian preadaptations, or exaptations, *are* new uses of screwdrivers.

So all that that has to happen in the exaptive evolution of a bacterium in a new environment is that some molecular screwdriver *finds a use* that enhances the fitness of that bacterium in that environment. Given heritable variation and natural selection that new *use*, hence new *function*, will emerge in the evolving biosphere. Patrick will, with his peptide stuck to a rock, become the first sessile feeder. But by the preceding discussion, we cannot prestate the new use of the screwdriver, so the new function is not prestatable. We do not know the sample space of biological evolution, whose becoming is therefore not a machine. We cannot prestate Darwinian preadaptations or exaptations, which are all co-optings of this for that. Further, the finding of a use that enhances fitness *is* the arrival of the fitter, a problem never solved by Darwin.

I suspect that we are beyond Gödel's theorem—which states, given a set of sufficiently rich axioms, there are formally undecidable statements from the axioms. If those statements are added as new axioms, there are new undecidable statements. I think the claims about screwdrivers go beyond Gödel who, after all, helped himself to an axiom set from which to formulate his theorem. There is no axiom set, for the evolution of the biosphere becomes willy-nilly, but contingently and not fully randomly. Life is not mathematizable in its specific becoming, and hopes for a prospective theory of the specifics of evolution are, I think, foredoomed.

Jury-Rigging and Whatnot

Figure 10.1 shows one of my favorite examples of jury-rigging. Ian Wilkerson, a colleague in Sydney, had a leak in his roof and asked a journeyman friend to help. The man rigged a funnel attached below the leak, to a tube leading out the front door over a railing, drooping toward the ground, slowly draining. Finding that a lamp in his house was hanging too low, the man also slung the lamp cord over the tube, jury rigging on jury rigging.

It all, well, worked until the real repair a few days later.

What is jury rigging? It is using a set of things or processes for purposes other than those for which they may have been designed to solve some problem.

We all jury-rig all the time. You may be interested to know the universal jury rigging kit: duct tape and WD-40 oil. If it moves, tape it. If it is stuck, hit it with WD-40! If desperate, rig a thingamajig out of the duct tape.

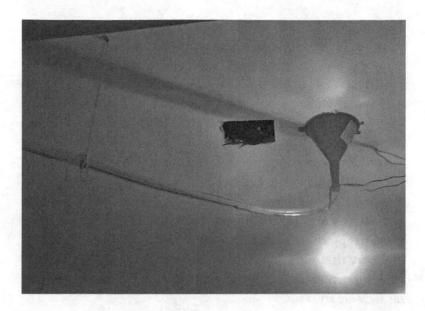

FIGURE 10.1 Jury-rigging: bricolage (Jacob 1977).

Could we have a deductive theory of jury-rigging? No, what would it be? The new uses to which we put things and processes to find an apt answer depend on the specific context of the leaking pipe or broken bike wheel. There is no deductive theory of jury-rigging to solve different problems, but we do it all the time. We are inventive. So is evolution, so was Patrick in particular. And none of us can predict in advance what we might invent and what might be invented from our inventions.

But there is something more to be said: Is it easier to jury-rig with lots of different things, a toolkit full of widgets and goop and string, and tape and springs, and so on? It is easier with lots of things.

The same is fundamental to the evolution of the biosphere. Exaptations are jury-rigged solutions to life's problems. The more diverse things and processes abound, the easier it is for creatures to jury-rig, well, at least something.

More stuff, more things, the more they are "fiddleable." The French have two expressions for this: *machinez le truc*, and *trucez le machine*. Thingamabob the whatdyacallit and whatdyacallit the thingamabob: this is F. Jacob's "bricolage" in evolution (Jacob 1977).

As Patrick begets Rupert, begets Sly, begets Gus, begets ... the more diverse creatures there are to interact with one another, the more ways become new adjacent possible opportunities for jury-rigged exaptations. And the more these exaptations create new creatures or creature features, the more they expand the total "context"; and thus, the more readily yet further exaptations can arise. In turn, this creates yet new creatures!

The biosphere explodes in diversity, creating more and more cracks in the floor of Darwin's nature until the cracks, ever expanding, become the very floor of nature, and nature herself.

Chapter 11

A World Beyond Physics

The aim of this chapter, indeed the driving purpose behind this book, it to show that life, though rooted in physics, surges beyond it into the myriad unprestatable ways of making a living in the world. Thanks to the three closures—constraint, work cycle, and catalytic—living systems literally construct themselves, and construct themselves upward into the unending openness of complexity in the nonergodic universe above the level of atoms. No laws describe or entail this miracle.

Entropy and Evolution

The famous second law of thermodynamics states that disorder, or entropy, increases in closed systems. Evolution is a story of increasing vast complexity and organization of organisms and ecosystems comprising the biosphere. Does the second law preclude the complex becoming of the biosphere? The answer is no. First, given open systems, the input of high-quality energy—for example, blue photons—allows thermodynamic work to be done, for example, in photosynthesis, with the release of red-shifted photons of lower energy. In this process, of course, entropy is produced.

But more than that, the union of the three closures—constraint closure and work cycle closure, plus catalytic closure—means that protocells and later cells literally do thermodynamic work to construct themselves, harnessing free energy available to them in doing so and producing entropy in doing so. Given heritable variation and natural selection of protocells and beyond, the creatures of the burgeoning biosphere build themselves upward into the complexity that they mutually create. They do so faster than the increase of entropy would degrade them. Order wins.

Niche Creation Is Self-Amplifying

We saw in chapter 10 that jury-rigging is easier the more tools there are in the tool kit. And we saw that much of evolution is due to Darwinian preadaptations, the unprestatable co-opting of organs and features for "this and that," like Patrick's peptide.

The increasing diversity of proto-organisms and organisms—Patrick, Rupert, Sly—creates ever more niches, which increases the diversity of "contexts," which increases the diversity of adjacent possible "uses," which in turn increases the ease of finding new ways to make a living in a biosphere exploding with possibility.

The filling of these niches by ever new, unprestatable organisms, creates yet further new contexts and opportunities. The total system "explodes" in a self-amplifying way into the very adjacent possible it itself creates. And, as noted, selection does not accomplish this magic of emergent becoming.

The same claims are true for the global economy, which has exploded in diversity from perhaps 1,000 goods and services—stone tools, for example, some 50,000 years ago—to billions today. Like species in a biosphere, goods and services afford niches for ever new goods and services, enabled to come into existence by what exists now. The mainframe computer of IBM did not cause, but, through the market it created, enabled the personal computers of Apple, along with the invention of the chip, and other makers; which did not cause but enabled word processing, spreadsheets, and companies like Microsoft; which did not cause but enabled modems and file sharing; which did not cause but enabled the World Wide Web; which did not cause but enabled selling on the Web with eBay and Amazon; which did not cause but enabled search engines such as Google. Each new good, starting with the personal computer, is a component enabled by the former. Strikingly, economic growth theory seems to ignore these facts.

In short, for the biosphere, and "econosphere," niche creation is self-amplifying. In both cases, the current system enables an unprestatable adjacent possible into which the system is "sucked." We become what it is next possible to become, and we ourselves create those very possibilities. The swim bladder creates the possibility that a worm could evolve to live in swim bladders.

This *is* life, explosively rich in its emergent complex, surging, unprestatable, and diversifying becoming—a myriad miracle of which we are part.

Beyond Law: Biology Cannot Be Reduced to Physics

As we saw in chapter 2, biology cannot be reduced to physics because physics cannot discriminate functions as subsets of causal consequences. The function of the heart is to pump blood, not make heart sounds. Further, the *only reason* in biology that such functions *exist* in the universe, hearts, for example, is that they abet the propagation and selection of the living forms of which they are parts. Hearts *only get to exist* in the nonergodic universe above the level of atoms because they are selected for the function of pumping blood sustaining the organisms of which they are parts. But we cannot deduce ab initio, 3.7 billion years ago, that hearts and swim bladders will emerge.

But there is more. We cannot even prestate the phase space of biological evolution.

In physics, one always prestates the phase space of a system. For Newton, given his three laws of motion, the phase space is defined by the boundary conditions, for example, the boundaries provided by a billiard table. Given these, we can define what we call the phase space of all possible positions and momenta—every way the balls can move on the table. Then we write Newton's laws in the form of differential equations; and from the initial and boundary conditions, we solve for the trajectories of the balls by integrating the equations.

Integrating Newton's equation is precisely to *deduce* the consequence of the differential equations for the trajectories of the balls, given the initial conditions and the boundary conditions. But deduction is logical entailment: all men are mortal, Socrates is a man, therefore Socrates is a mortal. Feel the logical force of the deduction.

What is true for the billiard table is true in general in classical physics. As Rosen (1991) said, Newton mathematized Aristotle's efficient cause as such deduction. The becoming of the Newtonian world machine is logically entailed by the initial conditions of the universe and Newton's laws.

But biology is different. Biological functions are part of the phase space of biological evolution: the trunks of elephants reaching for water, ears and middle ear bones and hearing, hearts pumping blood, swim bladders enabling sensing of neutral buoyancy in water columns.

But we cannot *prestate* the ever-changing phase space of ever new functionalities that arise! Therefore, we can write no laws or equations of motion for this emergence. And therefore, we cannot integrate the equations of motion that we do not have to yield entailing laws.

No Laws Entail the Becoming of the Biosphere

We can write no laws of motion, from the time of Patrick and Rupert, for the emergence of the eukaryotic cell, sex, multicelled organisms, the Cambrian explosion with its specific marvels of the explosion of diversity of early flora and fauna, promissory of us, fish, amphibians, reptiles, mammals, and primates, let alone the specific proteins that have emerged. We live in an unprestatable, literally unimaginable, myriad of emergent becoming. Because we can write no laws for the specific emergence we life, we are based on physics, but beyond physics.

The living world is not a machine, deducible by Laplace's demon for whom the world was deducible given Newton's laws and the current positions and momenta of all the particles.

Reductionism Fails

The biosphere is part of the universe. Reductionism, Weinberg's superb dream of a final theory, is of one that would allow us to deduce all that becomes in the universe—that would entail all. But no laws entail the becoming of the biosphere, and the biosphere is part of the universe, so reductionism fails. There is no final theory.

Thanks to the three closures—constraint, work, and catalytic—life literally constructs itself upward, the tree lunging for the sun. Life tweaks itself into the very cracks in Darwin's floor of nature that life itself creates for itself in its untellable fusillade of creativity. We have gotten from Patrick to a microbial world, a eukaryotic world, to a world of plants and animals, to Darwin's "forms most beautiful."

This vast emergent becoming is beyond physics, yet based on it. This is life co-constructing itself and enabling its own vast evolutionary diversification here, and on any biosphere, in the universe.

If among the 10^{22} solar systems estimated to exist, life is common, this self-constructing diversifying becoming is rife in the universe, is beyond physics, and may be as huge as physics in the emerging and growing complexity of the evolving universe as a whole.

This is a world beyond physics.

EPILOGUE

The Evolution of the Economy

Throughout this book, I have hinted at parallels between the evolution of the biosphere and the evolution of the economy. In this epilogue, I wish to develop those ideas. Fifty thousand years ago, the global economy may have had a diversity of a few thousand goods and services, including fire, unifacial stone scrapers, hides, and so forth. Today, in New York alone, there must be over a billion goods and services. The global economy has exploded in diversity. The question is how and why has this explosion occurred?

The economy, as detailed a bit further in the following, is a network of complements and substitutes, which I will call the "Economic Web." And like the biosphere, its evolution is substantially unprestatable, "context dependent," and creates its own growing "context" that subtends its "Adjacent Possible." The adjacent possible is what can arise next in this evolution. This evolution is "sucked into" the very Adjacent Possible opportunities it itself creates.

I do not wish to consider here the rich evolution of a single technology. Brian Arthur (2009) has brilliantly done so in his book, *The Nature of Technology*. Rather, I wish to discuss the evolution of the entire economic web; for as we shall see, goods and services create novel niches that invite the creation of new complementary and substitute goods such that the Web as a whole grows in diversity.

What Is an Economic Web?

The two central ideas are complements and substitutes. A screw and a screwdriver are used together to create value, such as screwing in a screw. And so they are complements. A screw and a nail can each be used to fasten two boards together. They are substitutes for each other. The economic web is the web of all goods and services; and for each, noted as a dot, a blue line connects it to all the other goods and services that are its complements and a red line connects it to all those other goods and services that are its substitutes. With billions of goods and services, this web is very complex indeed.

Two Senses of "Need"

In addition to goods and services are needs. A first sense of "need" for a good can be a need for its complement. A screw "needs" a screwdriver to be of use in screwing in a screw. A second sense of need is that we humans often need to fasten things together. Ultimately, the demand for a good or service depends on our purposes and needs. The latter is the basis of utility theory in economics. Utility theory tries to define, often mathematically, the trade-offs between goods from the point of view of a person, such as for consuming apples versus oranges.

Economic opportunities typically exist for unmet needs in both senses of the word. Economists typically focus on the second sense, but the first sense drives much of the evolution of the economic web, for a given technology *needs* its complements to be of use. So new technologies will drive economic growth by "needing" new complements. That need is an economic opportunity. In the second sense, we humans "need" word processing for its convenience in the preparation of documents. Thus, word processing emerged in the economic opportunity to fill that need, hence demand.

A Brief Look at the Evolution of the IT Industry

The world of information technology has exploded in the past 80 years. In the 1930s, Turing invented the Turing Machine, an abstract formulation of a digital computer. By mid–World War II, Turing's idea was crafted at the University of Pennsylvania into the ENIAC machine to calculate the trajectories of naval shells. After the war, von Neumann invented the mainframe computer; and shortly later, IBM made the first commercial machines, expecting to sell only a few. But the mainframe sold widely, and with the invention of the chip, paved the way for the personal computer.

Note that the mainframe did not *cause* the invention of the personal computer, but the wide market the mainframe created *enabled* the rather easy penetration of the personal computer into an expanding market. In addition, the spreadsheet is often described in histories of technologies as the killer app that caused an explosion of the personal computer market. The spreadsheet is the complement of the personal computer. Each helped the other gain market share.

The personal computer did not cause but enabled the invention of word processing, and software companies like Microsoft emerged, which was originally founded to make the operating system for IBM personal computers.

The invention of word processing and abundant files invited the possibility of file-sharing, and the modem was invented. The existence of file-sharing did not cause, but invited, the invention of the World Wide Web.

The existence of the Web did not cause, but enabled, selling on the Web; and eBay and Amazon emerged. And eBay and Amazon put content on the Web as did myriad other users, enabling the invention of web browsers; and so companies like Google emerged.

Thence has followed social media and Facebook.

Note now that almost all of these successive innovations are the complement of the preceding ones. The existing goods and services at each state are the "context" in which the next good and/or service emerges. Word processing is a complement of the personal computer, the modem a complement of word processing, the Web is a vast interconnected modem and is a complement and much more to file-sharing. The opportunity to share files "invited" the invention of the modem.

I note again that goods and services as contexts do not cause, but enable, the invention and introduction of the next good or service. "Enablement" is not a word used in physics.

A parallel history can be told of the automotive industry. The invention and advent of the automobile killed off the horse as a major mode of transport. With the horse went the smithy, the buggy, the buggy whip, and the corral. With the car came an oil and gas industry, paved roads, traffic control, motels, fast food restaurants, suburbia, and people living in suburbia needing cars to get to work in town. Gas is the complement of the car, the motel is the complement of the car, and so forth. Each stage in that evolution begets the next stage.

The Adjacent Possible of the Economic Web

Given the mainframe and personal computer, word processing is an opportunity in the adjacent possible of the economic web. The actual and the adjacent possible concern what exists now and what is next enabled to come to exist by the context of what is now actual. What comes next in the adjacent possible emerges out of what is here now, that is, the actual. In general, the next evolution of the economic web grows out of whatever is now actual and flows into the adjacent possible that the current actual enables.

The Algorithmic Adjacent Possible

Consider a Lego world. Start with a vast number of Lego blocks and place them on a central circle within a vast set of concentric circles like a target pattern. In ring 1, place all Lego objects that can be constructed from beginning Lego blocks in a single "legal" assembly, or "move," like moves in a kind of chess game. In ring 2, place all objects constructible in two steps, and so on to ring N out to infinity. The Lego block structures that exist "now" in, say, ring 7, unleash the adjacent possible of all the Lego structures that can next be constructed by single legal Lego moves.

This world is entirely "algorithmic" in the sense that there are "legal" Lego building moves. One may not, for example, use scotch tape to bind two blocks together instead of snapping them into place.

In a moment, we will see that the true adjacent possible of the economy is not algorithmic and not prestatable.

New Goods, Services, and Production Functions Can Arise as New Combinations

Consider the Wright Brothers' airplane. It is a combination of a light gas engine, airfoil, bicycle wheels, and a propeller. The printing press was a recombination of a wine press and movable type. New goods are often such combinations. For example, a parachute on the back of a Cessna could become an airbrake. Arthur (2009) makes the same point in *The Nature of Technology*.

Therefore, new technologies grow out of the technologies that now exist. The actual flows into its adjacent possible.

Thus, the economic web grows by creating its own "opportunities" to grow into the very adjacent possibilities it itself creates.

The Non-Algorithmic, Unprestatable, Adjacent Possible of the Economy

Lego world is algorithmic with legal and nonlegal moves of construction. The real economy is not so limited. In the main text of this book, I discussed the "screwdriver argument" and jury-rigging. I concluded that there is no algorithm that can list all the uses of a screwdriver, nor list the next use of a screwdriver.

But we find new uses for screwdrivers all the time. I need merely recall James Bond in a crisis using a screwdriver to turn the situation to his advantage.

But these new uses are typically unprestatable.

Moreover, these new uses are the very heart of innovation.

This is now being recognized by industry. For example, consider "crowdsourcing." Hey everyone, what is my new gadget useful for?

Thus, the innovative new uses of things and processes, enabled by what is now actual, is how the economic web expands unprestatably into its adjacent possible.

A charming true story exemplifies all of this. A few years ago, a man was living in Tokyo around the time the iPhone was introduced. He lived in a tiny apartment with a new baby, and it was crowded by his many books. He realized he could copy all his books with his iPhone then sell the books to create more space in his apartment. Then he realized his opportunity, rather like Patrick the protocell. Many other families in Tokyo lived in crowded apartments. He could go to these and offer to use his iPhone to copy their books, sell them, and take a percentage of the sale as his profit! His business was a success and is now itself being copied. What was his opportunity? Crowded apartments, the iPhone, and markets for books. The new business was his innovation.

We come to an important conclusion: the growth of the economic web is "sucked into" the very adjacent possible it itself creates!

The Unknowable "Size" of the Adjacent Possible

We cannot measure the "size" of the adjacent possible. We do not know what is in it. Consider flipping a fair coin 1,000 times and asking if it will come up heads 540 times. We do not know, but we can calculate the probability with the binomial theorem. We do not know what *will* happen, but we know what *can* happen. We know all 2 to the 1,000th power possible outcomes of flipping a fair coin 1,000 times. We know the sample space of the process.

But for the evolution of the economy into its adjacent possible, we do not know the sample space! Hence, we can construct no probability measure. So we cannot know the size of the adjacent possible.

The Diversity of Contexts and Diversity of Uses

The number of uses of a screwdriver depends on the diversity of the context. A screwdriver cannot be used to do much by itself in empty space, but it can be used to do many things alone or with other things in New York in 2017.

In the main text, I briefly considered jury-rigging. I concluded that there can be no deductive theory of jury-rigging. But it seems we can say something. If you confront some arbitrary problem, would you be better off jury-rigging with a single object or process, say a screwdriver, or an assembly of many objects: screwdriver, duct tape, a shoe horn, old battery, bailing wire, nails, a sheet of cloth . . . ?

Obviously, it is easier to jury-rig with lots of objects lying at hand than a single object. While we seem unable to quantify this, at least now, it seems clearly true.

In short, the diversity of "context," here the number of objects available, is related to the number of "things" one can do with the assembly. A garage full of stuff is more easily commandeered to new ends than a clean one.

The Growing Web *Is* the Growing Context for Its Own Further Growth

As new goods and services and production capacities come into existence, they provide the growing contexts into which yet more new goods and services and production capacities can follow as their complements or substitutes. An economy with a high diversity of goods, services, and production functions is rather like a garage full of "stuff" rather than a clean garage. It is easier to jury-rig in the garage full of stuff, and it is easier to invent new goods and services and production functions in an economy already full of such stuff. But the new goods, services, and production functions only make the "garage" more full of stuff; thus, amazingly, the economy grows its own adjacent possible and *augments* that very growth as the growth occurs. The process is broadly self-accelerating.

Thus, the growing economic web explodes in a diversity of complements and substitutes from perhaps a thousand or ten thousand goods 50,000 years ago to billions today!

But the same holds true for the evolving biosphere, as seen in the main text of this book—from Patrick, Rupert, Sly, and Gus— to the expanding diversity of species in the Phanerozoic aeon in the last 600 million years. New species literally create niches for yet further new species. New goods create niches for yet further new goods and services and production capacities.

A Brief Comment on Standard Economic Growth Models

What I have sketched here is very different from most standard economic growth models. These model the economy, not as a web, but a single sector making, in effect, a single product. Then one considers input factors such as capital and labor and human knowledge, investment, and saving and writes differential equations that can model growth. These work fine to some extent, but not for an economy that is creating ever new goods and services, as is our economic web.

An Early Statistical Model of the Adjacent Possible

We have, at present, no mathematical model for the unprestatable evolution I describe previously. However S. Strogatz and V. Loreto (Loreto et al. 2016) have taken an important first step. Theirs is a first model of the adjacent possible. They start with what is known in mathematics as a Pólya urn model. In this model, one starts with an urn holding 50 percent black balls and 50 percent white balls. The player randomly picks a ball. If it is white (or black), he replaces the ball and adds one white (or black) ball. The question is, after a long time, what is the stationary fraction of white balls? The answer is "any value between 0 percent and 100 percent equiprobably. That is, one could have 69 percent black balls and 31 percent white balls, or 0 percent black balls and 100 percent white balls.

In Strogatz and Loreto's (Loreto et al. 2016) variation, one starts with at least two colors of balls. All picked balls are replaced in the urn. But if a color is picked that was not seen before, it is replaced, and a random *new* colored ball is introduced. The new color models the new adjacent possible. The process continues indefinitely. The process generates a power law distribution of colors and fits both Zipf's law and Heap's law. The random new colors are an initial move to model the unknowable adjacent possible. Their fit to both Zipf's law and Heap's law with much data is encouraging.

The model is lovely, but does not yet answer our needs, for it is one of a branching set of independent lineages of descendant colored balls. A red ball gives rise to an orange ball, which gives rise to a blue ball. There is no cross-talk between lineages augmenting the combinatorial formation of new colors as there is in the economic web's evolution with new complements and substitutes arising from old ones by new jury-rigged combinations of one or several prior goods. I hope that a good model or set of models can be constructed.

This epilogue extends the ideas of the main text about the unentailed evolution of the biosphere as species create niches for one another, often adapting by unprestatable Darwinian preadaptations into the expanding adjacent possible of the evolving biosphere to what seems to be much the same set of processes in the evolution of the economy. In both cases, like the garage filling up with ever newly invented "stuff" for jury-rigging, life creates its own staggering possibilities of future becoming.

To think that this is a Newtonian-Laplacian machine, derivable in its specific becoming from some set of axioms, seems deeply wrong. Life, and we among it, is so rich in its inheritance and prospects that we can, I think, be captured by no entailing laws.

REFERENCES

Arthur, Brian W. (2009). *The Nature of Technology*. New York: Free Press.

Atkins, Peter W. (1984). *The Second Law*. New York: W. H. Freeman and Co.

Damer, B. (2016). "A Field Trip to the Archaean in Search of Darwin's Warm Little Pond." *Life* 6: 21.

Damer, B. and D. Deamer. (2015). "Coupled Phases and Combinatorial Selection in Fluctuating Hydrothermal Pools: A Scenario to Guide Experimental Approaches to the Origin of Cellular Life." *Life* 5, no. 1: 872–887. https://doi.org/10.3390/life5010872.

Dawkins, Richard. (1976). *The Selfish Gene*. Oxford, UK: Oxford University Press.

Djokic, T., M. J. Van Kranendonk, K. A. Campbell, M. R. Walter, and C. R. Ward. (2017). "Earliest Signs of Life on Land Preserved in ca. 3.5 GA Hot Spring Deposits." *Nature Communications* 8: 15263.

Dyson, Freeman. (1999). *The Origins of Life*. Cambridge, England: Cambridge University Press.

Erdős, P. and Rényi, A. (1960). *On the Evolution of Random Graphs*. Hungary: Institute of Mathematics Hungarian Academy of Sciences Publication, 5.

Farmer, J. D., S. A. Kauffman, and N. H. Packard. (1986). "Autocatalytic Replication of Polymers." *Physica D: Nonlinear Phenomena* 2: 50–67.

Fernando, C., V. Vasas, M. Santos, S. Kauffman, and E. Szathmary (2012). "Spontaneous Formation and Evolution of Autocatalytic Sets within Compartments." *Biology Direct* 7: 1.

Hordijk, W. and M. Steel. (2004). "Detecting Autocataltyic, Self-Sustaining Sets in Chemical Reaction Systems." *Journal of Theoretical Biology* 227: 451–461.

Hordijk, W. and M. Steel. (2017). "Chasing the Tail: The Emergence of Autocatalytic Networks." *BioSystems* 152: 1–10.

Jacob, Francois. (1977). "Evolution and Tinkering." *Science New Series* 196(4295): 1161–1166.

Kauffman, S. A. (1971). "Cellular Homeostasis, Epigenesis, and Replication in Randomly Aggregated Macromolecular Systems." *Journal of Cybernetics* 1: 71–96.

Kauffman, S. A. (1986). "Autocatalytic Sets of Proteins." *Journal of Theoretical Biology* 119: 1–24.

Kauffman, Stuart. (1993). *The Origins of Order: Self-Organization and Selection in Evolution*. New York: Oxford University Press.

Kauffman, Stuart. (2000). *Investigations*. New York: Oxford University Press.

LaBean, Thomas. (1994). PhD thesis, University of Pennsylvania Department of Biochemistry and Biophysics.

Lincoln, T. A. and G. F. Joyce. (2009). "Self-Sustained Replication of an RNA Enzyme." *Science* 323: 1229–1232.

Longo, G. and M. Montévil. (2014). *Perspectives on Organisms: Biological Time, Symmetries and Singularities*. Berlin: Springer.

Longo, G., M. Montévil, and S. Kauffman. (2012). "No Entailing Laws, But Enablement in the Evolution of the Biosphere." In *Proceedings of the 14th Annual Conference Companion on Genetic and Evolutionary Computation*, 1379–1392. See also http://dl.acm.org/citation.cfm?id=2330163.

Loreto, V., V. Servedio, S. Strogatz, and F. Tria. (2016). "Dynamics on Expanding Spaces: Modeling the Emergence of Novelties." In *Creativity and Universality in Language, Lecture Notes in Morphogenesis*, edited by M. Degli Esosti et al. Basel, Switzerland: Springer International Publishing.

Montévil, Maël and Matteo Mossio. (2015). "Biological Organisation as Closure of Constraints." *Journal of Theoretical Biology* 372: 179–191. http://dx.doi.org/10.1016/j.jtbi.2015.02.029

Prigogine, Ilya and Gregoire Nicolis. (1977). *Self-Organization in Non-Equilibrium Systems*. New York: Wiley.

Pross, Addy. (2012). *What Is Life? How Chemistry Becomes Biology*. Oxford, England: Oxford University Press.

Rosen, Robert. (1991). *Life Itself.* New York: Columbia University Press.

Schrödinger, Erwin. (1944). *What Is Life?: Mind and Matter?* Cambridge, England: Cambridge University Press.

Segre, D., D. Ben-Eli, and D. Lancet. (2001). "Compositional Genomes: Prebiotic Information Transfer in Mutually Catalytic Noncovalent Assemblies." *Proceedings of the National Academy of Sciences USA* 97: 219–230.

Serra, Roberto and Marco Villani. (2017). *Modelling Protocells: The Emergent Synchronization of Reproduction and Molecular Replication.* Dordrecht, The Netherlands: Springer.

Snow, Charles Percy. (1959). *The Two Cultures.* London: Cambridge University Press.

Sousa, F. L., W. Hordijk, M. Steel, and W. F. Martin. (2015). "Autocatalytic Sets in E. coli Metabolism." *Journal of Systems Chemistry* 6: 4.

Vaidya, N., M. L. Madapat, I. A. Chen, R. Xulvi-Brunet, E. J. Hayden, and N. Lehman. (2012). "Spontaneous Network Formation Among Cooperative RNA Replicators." *Nature* 491: 72–77. doi 10.1038/nature11549.

von Kiedrowski, G. (1986). "A Self-Replicating Hexadesoxynucleotide." *Angewandte Chemie International Edition in English* 25, no 10: 932–935.

Wagner, N. and Gonen Ashkenasy. (2009). "Systems Chemistry: Logic Gates, Arithmetic Units, and Network Motifs in Small Networks." *Chemistry: A European Journal* 15, no. 7: 1765–1775.

Weinberg, Stephen. (1992). *Dreams of a Final Theory.* New York, NY: Vintage Books.

Woese, C. and G. Fox. (1977). "Phylogenetic Structure of the Prokaryotic Domain: The Primary Kingdoms." *Proceedings of the National Academy of Sciences USA* 74: 5088–5090.

INDEX

Page references followed by *f* and *t* indicate figures and tables, respectively.